Beyond Darwinism, Waiting at the Gate of Eden

Beyond Darwinism Waiting at the Gate of Eden

A memoir about revisiting human evolution through ancient creation myths

To John,
my wonderful
Teacher

Farideh Shadravan

Bahman Press
San Francisco

Library of Congress Control Number: 2007933783

ISBN: 978-0-615-15356-8 Paperback

To the ideals of equality for

all humankind despite its divided path

Contents

Keynote

Our human genome and fossil bones point to a long organic evolutionary trail behind us, whereas our myths are full of narratives about our divine origin. The answer to "Who are we?" may be embedded in another question. Do the human mind and body have the same or separate origins?

This book is partly about unfolding the saga of human evolution, partly about retelling the forgotten folk tales of human creation. It is also about the life experience of a scientist prepared to view these two parallel events from a single vantage point. Written particularly for all those interested in the origin of humanity, it is from an evolutionary geneticist's perspective, which is tainted with a shocking twist at the end. Arguably, the source of this apparent anomaly could be traced to the reality that it is written by someone who came to the US at age 22, learned her craft according to Western traditions, however, still managed to perceive the world around her like most Middle Easterners do.

In this book, to better understand the origin of human mind, I let the East and West meet in my own mind. One outcome of this fateful encounter, albeit utterly unacceptable to our contemporary mindset, may ultimately bear out to be a more viable choice in thinking who we are than the alternative views long advocated by powerful scientific and religious establishments. Ironically, this unusual view is in line both with well-known genetic principles and the narratives of ancient religions. But, it inherently rejects the current consensus that *Homo sapiens* and Adam both are eligible to be regarded as our primal ancestors. Unless we take a giant leap in our basic assumptions, as tentatively suggested in this book, we could never bring these two presumed ancestors of ours to meet each other face-to-face.

Understandably, there are always some intensely spiritual people who abhor any notion of mixing spirituality with physical entities, and likewise there are geneticists who cannot envision any place for spirituality within DNA premises. However the riddle of human existence would be best approached by allowing ourselves to view the human body and soul under the same light.

Introduction

A while ago, I read something quite astonishing about Mandaeans, a little known quiet people living on the tip of the Persian Gulf, in southern Iran and Iraq. They are the strict and pious followers of an ancient Gnostic tradition believing in the teachings of several old-time prophets (see p. 218). However, there is something quite unusual about their religious practices. Although, their symbol is a simple and bare cross, they are not followers of Jesus Christ. They say, he revealed too much.

It is very ironic, but we humans do have bizarre ways of dealing with knowledge and how we acquire it. On one hand, we have founded our human civilizations based on the merits of knowledge, and on the other hand, we have always, since antiquity, shunned those who try to delve in it too deeply. But make no mistake about it; this is not an exclusively Eastern syndrome. While the Easterners tend to suppress what they do not agree with, the Westerners indirectly show their intolerance, by persuading people, through illusive measures, to join them and follow their path.

This book is not about why we humans behave as such; rather, it is about attempting to unravel some secrets about the source of human knowledge. It is about pondering the origin of the human mind. Our mind, as part of the human body, could have had an organic origin, a product or byproduct of various evolutionary forces acting together. But at the same time, there are so many unrelenting ancient myths, still in our everyday thoughts, persisting in amusing us, in provoking us, no matter how hard our current scientific and religious establishments persuade us to abandon them for good.

Four years ago on a rainy afternoon in November 2002, the start of rainy season in San Francisco, I took BART (Metro train) to go downtown. Soon I was talking to a passenger sitting next to me. She was an accountant between jobs, maybe a few years younger than me. I was a geneticist who had left my job in a big biotech company a few months earlier and was pondering whether to go back and find another job as a scientist; or after some 30 years in science, both as a graduate student and a professional, carve out a non-scientific career for myself.

We talked for a while and after she learned that I was a geneticist, her eyes twinkled like two bright stars in the dark midnight sky and with an enchanting voice she said, "You must be a genius." I was utterly stunned by her quick conclusion. Instead of savoring the moment, however, I felt rather embarrassed. Embarrassed for the enthusiasm of non-scientific people who have so overwhelmingly and unconditionally embraced the recent advances in genetic research; and embarrassed for geneticists themselves who have brought this euphoric sense of "mission accomplished" to the naïve and beguiling general public.

Only if non-scientific people had worked in a scientific field for a couple of years, they would have known better to curtail their enthusiasm and react more cautiously to everyday scientific news published in newspapers. Whereas most people usually carry out their routine jobs in more or less factual situations in regular buildings, we scientists dwell in ivory towers founded, in principle, on assumptions. Thoughtful and reasonable assumptions based on inferences, inductions, deductions, conclusions and so forth.

One further hindrance in this world of guesswork and propositions is that not all assumptions are equally acceptable to the elitist mind of scientific establishment. Ten years ago, I learned from my own experience that like other establishments, the first and foremost priority of the scientific establishment is to safeguard its hard fought values and principles. After four years of independent research, analyzing numeric data some16 hours a day, writing and rewriting eight scientific papers, based extensively on statistical analysis, none of my papers could ever see the light of day and any chance of publication or even peer review. Unfortunately, my ideas and my data did not fit within the accepted boundaries of the current mindset of genetic research. As frequently reminded by some of my close associates, I was way ahead of my time. That is, my ideas have to wait for a new genetic paradigm to take shape in the unknown future. Easier said than comprehended why a scientist has to wait at all.

About a century has passed since Thomas Hunt Morgan and his wife Lillian, two early geneticists, started their research with tiny fruit flies, the same organisms I spent all my graduate years working with. These miniature flies, only a few millimeters long, have big and beautiful eyes usually in bright red, and also in a variety of other colors such as white, pink, purple, apricot, ruby, brown and many more. The variation in the color of their eyes, along with some other physical traits, inspired the early geneticists to develop some important theories about the laws of inheritance.

Since then, hundreds of thousands of scientific papers have been published. If only a fraction of the assumptions presented in these papers were accurate, by now different forms of cancer would be fully curable. This statement is by no means to diminish all the great achievements genetic research has attained during the last century. But as a geneticist, I feel we have not accomplished enough to deserve the current full public trust in anything and everything genetics. There have been real advances benefiting the common people on the street, such as those in the field of human genetic identification, which have proven invaluable in criminal investigation and paternity cases involving runaway fathers.

And then there are some presumed advances, like those concerning our understanding of human evolutionary history, which I would rather call genetic "fairy tales." My favorite genetic tale has to be a recent attempt to put together substitutes for the mythical Adam and Eve. According to a hasty interpretation of a very limited DNA data, Adam and Eve were located at specific locations in the distant past, albeit tens of thousands of years apart from each other (see p. 165). As an evolutionary geneticist, I wonder why not disclaim the myth of Adam and Eve altogether, instead of trying to sabotage it so ruthlessly.

Are they trying to assure the public that genetic field has advanced so much that now it can provide all the details of human's prehistoric past? Are they implying that there should be no further need for the masses to entertain themselves with their old myths about human creation and other prehistoric events? One can only hope that this is just another acute case of scientific arrogance and not an attempt to suppress other views about humanity's past. It reminds me of Paul Feyerabend, a philosopher of science, who advocated the separation of science and state. I hope we do not have to go that far.

In principle, scientific assumptions are made to be readjusted constantly with more data or new sets of data. But sometimes under certain circumstances, some scientists tend to underestimate the fragility of the whole scientific enterprise, and start believing their own assumptions, and pontificating in the name of science. Currently, after some thirty years of great advances in molecular technologies, particularly in the DNA field, some geneticists are at that stage of grandiose illusion about our true understanding of genetic principles. In fact, during the past three decades, our genetic understanding has been modest, and only three major concepts have been generally accepted to be part of the genetic knowledge base: transposable elements, epigenetics, and micro RNA interferences.

There is a profound difference between the impact of technical and conceptual advances, and in the long term, it is the conceptual ones that matter most. For example, the first draft of the human DNA sequence, which was announced in 2000, represents a great milestone in technical advances, but so far it has made very modest conceptual contributions.

So considering that scientific reasoning, at times, could be at the mercy of circumstances, turning assumptions into short-term realities, I wonder why not allow other circumstantial practices of the human mind such as revelation and other mystical experiences enter the arena of organized practice of discovery, giving it a chance to become gradually mainstream. Obviously with our contemporary mindset, this proposition sounds like pure nonsense. However, after reading this book, hopefully the Western readers will have a better understanding of the extent of similarities between some genetic principles, particularly those relating to how the DNA functions and many West Asian early concepts relating to deities and other supernatural beings.

This book is about my mental journey from a rational geneticist to an irrational speculator pondering about a new and unorthodox view of the origin of modern humans. Thanks to this new idea that little by little grew in my head like an unwanted weed, now I realize how easy it is to be a rational scientist and to follow the scientific path laid down by other scientists before you. Now I realize why most people want to be rational and why societies encourage their citizens to be rational. It is so easy on everybody, and makes life more prosperous for all participants.

Nevertheless, sometimes during our lives, some of us reach a point where we have to respond to unknown forces within us to get out of our comfort zone and follow new directions opening in front of us. Four years ago, after 30 years in science, I found myself at that crossroads between rational and irrational thinking.

Surprisingly, in the first decade of the 21st century, there are many people well positioned to become more and more irrational thinkers. For example, many young people, with their minds preschooled in the futuristic adventures of humans in faraway galaxies, are ready to escape the gravitational force of reasoning and leap into other dimensions of the human mind.

Furthermore, with widespread use of the Internet at a global level, rigid textbooks are being challenged by the inconsistent virtual world of the Internet. Indeed, these days, as far as knowledge goes, it is getting very chaotic out there, and that is exactly what is needed for the development of a new mental dimension for *Homo sapiens*, the wise human.

PART ONE

LOOKING BACK

Behind every observation, there is an observer with a unique worldview, reflecting a particular life experience. Therefore, in evaluating an observation, its observer should be identified beforehand. A brief account of the scientific and cultural backgrounds of the observer in this book is given in the first part.

Working For Our DNA Heritage

Pictures at an Exhibition

The great ballroom of the Beaurivage Hotel was shimmering under a huge chandelier hanging from its majestic ceiling. We were having a banquet lunch at the Eleventh International Human Identification Conference in Biloxi, Mississippi, a resort town down the Gulf Coast. The room was filled with excitement with over 1,000 international attendees. While having a fancy launch at a large round table, I was talking with an English scientist, whom I had just met at my poster session. Then, an unassuming petite woman in her mid-fifties with a nice short haircut and a black jacket over a long dress approached our table and quietly sat next to me on the only empty chair available at the table.

The English guy looked at her respectfully and while pointing to me, with his soft voice told her, "This is Farideh." Suddenly the woman almost jumped out of her seat and enthusiastically said, "I wanted so much to meet you." Completely surprised, I asked her rather indifferently but with a polite manner, "Who are you?" She replied, "I am Julie." I asked,

"Julie who?" And then after she gave her family name, it was my turn to get really excited, "You are the Julie…." Indeed she was a big shot, unlike me.

She was not a scientist, but the head of Forensic Science Services (FSS), a government owned company that is the principal provider of forensic science to the UK criminal justice system. FSS was the biggest customer of our DNA forensic kits. A customer so special that we had developed a special DNA kit for them to be used in the UK. There, a somewhat different set of DNA markers was being used for DNA forensic work to identify criminals.

At the time, I was working in the Human Identification Department of a big biotech company located south of San Francisco. Although we had developed various DNA kits and reagents for DNA sequencing and DNA fragment analysis (as used for forensic works, among other applications), we were world-renowned for our automated DNA sequencing instruments. Our newest instrument was a completely automated high throughput machine called 3700. Its output and automation were much higher than those of our previous models operating in most forensic labs all over the world, albeit they were just a few years older models.

FSS in the UK was planning to replace its fleet of older DNA sequencers with 10 new 3700 machines, with the price tag of half a million dollars a piece. Celera Genomics, an offshoot of our company, had just finished the first rough draft of the human DNA sequence a few months earlier, using 500 of our 3700 machines supplied by its parent company.

I still vividly remember the time that our application department was testing each 3700 machine one by one before sending them to Celera. It was such an exciting time and we were learning so many new things every day about all the little problems that could interfere with the smooth operation of this superbly complicated machine, with its super slim structure, which had already won a prestigious prize for industrial design. Employing these super machines was one of the major factors with which Celera could achieve the historic feat of sequencing the human DNA, well ahead of time, and win the race over a publicly founded consortium of several DNA sequencing centers across the US. The achievement announced in the White House in June 2000 had created a great buzz among large DNA sequencing centers and everyone associated with this field was curious about the 3700, and so was FSS in the UK.

But FSS was another story. They would not just send one of their people to check the machine at our company, as was customarily done by other places. Their practice was to set up a six-month project involving several personnel in one of their labs to test a few of the machines of their interest, before actually committing themselves to purchase them. So we had loaned them three 3700 for extensive testing just to keep a good customer happy. Our real business interest in dealing with FSS, however, was to sell them our expensive DNA forensic kits worth several million dollars per year.

But in this case, our regular application specialist at FSS, a competent and professional English girl living there, could not be of much help to them to tease out their problems with this new machine. The reason was that the 3700 was technically just a different beast than our previous DNA sequencers. We, in our department, were well aware of their problems, but thought they were just being too picky, as usual. Until one day when by chance, I met a guy from FSS in a meeting in Washington DC, and the next thing I knew was my boss calling me to the conference room and asking me, "Farideh, would you like to go to the UK for a week?"

I was an R&D (research and development) scientist, and had nothing to do with the application group in charge of handling customers' problems. But apparently, after talking to that English fellow in Washington DC for about half an hour about the 3700, he thought I was a 3700 guru. So now FSS had demanded from our department that I go there and fix their problems. Well, life could not have been any fancier. I went to the UK on a business class flight, and came back on first class; was housed in a fancy little hotel with old English charm, spooky long narrow halls covered with old oval mirrors on their walls, enchanting real Victorian restaurants, and the best breakfast service anywhere in the world. All expenses paid, courtesy of the company. No wonder our 3700 machine had a half million-dollar price tag. I went to work at FSS every morning in one of those roomy black cabbies seen in old English movies. Upon arrival, somebody was waiting for me at front door to take me upstairs to our lab, as the ten-story large building was off limits to the public.

Apparently, they liked me so much there that, after four days, one very late afternoon, while I was still working on one of the machines, the head of the whole DNA forensic section, whom I had not met before, came to our lab looking very tired after a seemingly long day. He sat on a high chair, lowered his face to my level and seriously asked me, "Would you like to work for us here permanently? Can we hire you?" Taken completely by surprise, I stared at him for a moment, then without thinking at all, I said, "Sorry, I have to go back." While disappointed by my answer, he continued, "So at least can you stay here longer? How about two weeks?" Then he quickly added, "If you stay longer, I will personally take you to a nearby castle this weekend and show you around." My, my, my, my, was I hot or what? It seemed they were really desperate to finish their 3700 project on time. I said, "I would love to stay longer, but first you should talk to my boss."

So I stayed there for two weeks, and when I came back to the biotech company, my routine problems with our wonderful marketing group got even worse. They were not used to seeing a foreigner, God knows from where, outshine them like that.

Now, four months later, at the banquet lunch at the Beaurivage Hotel, talking to Julia, the head of FSS, she told me enthusiastically, "Now everybody knows you at FSS," adding that the reason she wanted so much to meet me was to know what I was thinking about the 3700 machine that they were still in the process of purchasing. After hearing my opinion, she looked me in the eye and whispered, "I think you are in love with this machine." What an amusing statement. I was in love with a machine. Strangely enough, I had heard that comment before, albeit in a different setting.

A few months earlier, a group of scientists from the DNA forensic section of the FBI, in Quantico, Virginia, were visiting our campus to check on the 3700, among other things. Although it was not part of my regular job, my boss came to our R&D lab, took me out of there, and asked me to display the 3700 for them. After I finished my presentation, Bill, the head of the FBI group, a small framed middle-aged guy with curious little eyes, who reminded me of Dr. Zachary Smith from the old TV series *Lost in Space*, smiled and announced to everybody, "I have never seen anybody talking so enthusiastically about a machine." But the 3700 was not just any machine. It was a half a million-dollar miracle maker. And compared to very early DNA sequencing instruments, it deserved every accolade I was bestowing on it.

About two decades ago, in 1983, when I started to do DNA sequencing, we had to spend a lot of time just to get the sequence of a few hundred lousy nucleotides per week. Those days DNA sequencing was performed in a very thin slab gel sandwiched between two large glass plates. We also had to use radioactive materials to label sequencing reagents, and run the instrument under high voltage inside a lab hood in case any accident happened. I remember one day, when I was running a gel, one of the glass plates suddenly broke. It was due to high temperature, and radioactive buffer poured all over the counter under the hood. Of course we were all well prepared for such likely accidents, but so much of our precious time had to be wasted on the elaborate clean up process.

In spite of my great enthusiasm for the technical advances we were making to further facilitate the DNA research, I could not last long in that company. About 18 months after the Beaurivage meeting, I left it for good. The best part of being there was doing the actual work, and the worst part was dealing with some of my supervisors and coworkers. Before joining that biotech company, I had spent over 20 years studying and working in seven universities across the US. During that long period, I never felt any need to reinvent myself to survive everyday routines. But corporate culture was different.

It was more like being in an obedience school for dogs. The funny thing about it, however, was that not only headmasters but also other dogs would take part in the training process. At first, I did not notice the pattern, and took it personally. So everyday became like a judgment day for me, re-evaluating some aspects of my cultural values that were set long time ago. I remember once I commended a friend for her nice skirt, and told her that she should wear skirts more often. But she was quick to say, "No, no, today I have to go somewhere special, otherwise, since my supervisor is a woman, it won't be wise to irritate her by dressing better than she does." And here I was, always dressed up, as if going to give a talk in a conference.

Dressing nicely, however, for me, was part of being Iranian. When I was growing up in Tehran in the sixties and mid-seventies, it was so common to take care of one's appearance. My father used to go to Europe every May for the Hanover International Industrial Show in Germany, and would always bring back beautiful Italian dresses for us. Besides, we Iranians as infants open our eyes crawling on Persian rugs, masterpieces of harmony in colors and design. T-shirts and pants just would not do it for us. They are so boring to wear day after day.

In our big company, from how to dress to how to talk, every aspect of our behavior had to be adjusted constantly so as not to upset the regimented spirit of these establishments. Of course, it was fine for marketing people to be outspoken and dress as they pleased. But every newcomer to our company would soon learn that R&D people should act like second-class citizens. After all, we were paid very well, at least compared to what we were used to in the universities. So what was the problem? The problem was that while we were all a bunch of women there, most R&D people had PhDs, whereas most marketing people had only BS degrees. The problem was that marketing people were mostly American or European women with no funny accent while R&D women were mostly from third world (now called nonaligned) countries, such as China and India.

I remember on many occasions, after our regular meetings with marketing people, my supervisor, a European immigrant, would take me aside and remind me that I should not argue with the marketing people. But it turned out that keeping quiet was not enough for him. Next, he warned me that I should be more careful with my eye movements, that is, the way I looked at the marketing people. The day I heard that bizarre warning, I told myself, "I am out of here." No matter how well they paid us, or how much they pampered us occasionally, this was not right. According to them, what I was supposed to do was shut up and keep my eyes fixed at the gray carpet floor, dressed in a uniform to survive this modern day cemented jungle.

I wondered what ever happened to the egalitarian spirit of Silicone Valley. I suppose that was the good old days in small startup companies where R&D people were in the driver seat. Back here at our company, we were more like working in "China West Incorporated." All for the sake of the company's bottom line. Soon after that demeaning warning, I decided not to report to my daily boot camp anymore. That was exactly how I felt everyday going there. I remember, after finishing the work every evening, when turning off the light under the overhead cabinet in my cubicle, I was always wishing that it was my last day there. And at last one day it was.

Although by nature I was a quiet and patient type, not believing in showing off for easy rewards, I was never faint at heart. I could stand my ground. Right or wrong, I believed that for every action, there should be a reaction of equal size in the opposite direction, as the first law of physics implies. But even as a die-hard fighter, I knew very well that not all battles are worth fighting for. Sometimes one has to leave the meek, the opportunist, and the aggressor behind to feed on their misery and gradually annihilate each other one by one; to hell with them all.

The day I left that company, driving home towards the foggy sky of San Francisco in the horizon, with teary eyes, I remembered vividly yet another painful exit in my career's colorful history file. It was the day I got a phone call that Sara, the head of the Biological Sciences Department at Lawrence Berkeley Lab (LBL), a national defense laboratory administrated by UC Berkeley, wanted to see me. It was December 6, 1994, and I went to her office the same afternoon.

She was a petite woman in her mid-fifties. I knew that she got her PhD in cell biology from Harvard, and did her postdoctoral work in the lab of a very well-known scientist at UC Berkeley. So she had a perfect scientific pedigree, which is very important in the aristocratic world of academia. She had a big lab at LBL, doing research on breast cancer. Although she was Iranian like me, with an accent just like mine, and I had seen her many times in our departmental meetings during the last three years, I had never got a chance to talk to her before. When I entered her office, she seemed kind of busy. So I was expecting a short meeting. As soon as she saw me, she said, "What is going on here?" I replied rather confidently, "Could you first tell me what have you heard? And then I will tell you my side of the story." She said, "Tim called me the other day and said that Farideh has vanished with her lab notebooks." I thought, very intriguing indeed. So I said, "I have not vanished. I officially quit my job with one month prior notice, and I have left all my lab notebooks in the lab." Then I opened a small binder in front of me, and took all its ten loose pages out, and pushed them respectfully towards her across the table, adding, "This is what all this commotion is about." And continued by saying, "I have already entered all this information in our official computer database. These sheets are just my scrap notes before entering the data into the database."

Apparently, Tim and other people in his group were acting as if they were afraid that I might use those data to set up a startup biotech company, as it was very common in those days. But they knew very well that my real interest in science was not in the established fields; rather, I was interested in the uncharted territories of evolutionary genetics. Tim, about the same age as Sara, was originally a physicist. Then he got interested in cancer research and switched to UCSF (University of California in San Francisco) and established there a large interdisciplinary research group. Meanwhile, at LBL Genome Center, I was working with a large genomic library for mapping human chromosome 21, as part of a big group. One day, he came and persuaded me to join his group to work on a joint project between LBL and UCSF.

The project was about mapping hundreds of DNA probes associated with various human diseases, using a mapping technique called in situ hybridization. With this method, by applying specific fluorescent dyes, we could localize each DNA probe to its specific place on each human chromosome, and then actually visualize each of them using a specialized microscope. Although moving the slides containing these probes under the microscope can give you motion sickness after a few hours, the images of stained chromosomes are absolutely breathtaking. I used to call them waltz of the chromosomes.

There were several people involved in this project. My job was to find each specific DNA probe in a genomic library containing about 140,000 clones, using an ultra fast and very innovative screening method. Today, with the advent of commercial microarrays, this screening method seems rather cumbersome and laborious, but in 1993, it was indeed considered state-of-the-art in DNA technology. With accurate screening, it was possible to find out the exact genomic address of each DNA probe in the huge genomic library rather quickly.

The next step was to go to the actual genomic library, arrayed in numerous special plates and housed in two large freezers in our building. Each plate contained 96 little wells, and each well was filled with a few drops of liquid holding a very small piece of DNA, part of the genome of a specific person. Actually, in the case of this library, to have a full coverage of every fragment of human genome, a mixture of DNA from seven people was used. Generally speaking, a genomic library contains all the DNA of one (or more) person(s) whose DNA is cut into very small pieces and cloned (inserted) into yeast or bacterium for easy propagation.

After locating the exact genomic address of the DNA probe of interest, that is, the position of a certain well in a specified plate, a tiny amount of liquid would be taken out from this well, and after some further preparation, it would be sent to another lab to be localized, by in situ hybridization, to its specific place on human chromosomes.

The promise of this publicly funded project was to enable many researchers, inside or outside the US, to do detailed genomic study on any particular disease, provided that the gene for the disease was within the DNA probe of their interest. Each DNA probe was like a book, ready to be studied, from a huge genomic library. Our job was to send them the right probe, and their job was, by using positional cloning or other methods, to narrow down the exact position of the disease gene of their interest to a unique DNA sequence on a specific probe.

After working on this project for about two years, and isolating several hundred DNA probes, I had to quit my job out of frustration. I found it very discouraging to do all the work and then see other people get credit for it. The day I left that job, I was warned by one of Tim's protégés, who was my direct supervisor, that he would persuade Tim to ask the LBL to sue me. So just before I left, I had to yell at him saying, "Don't mess with me, because I am not a quitter." Ironically, up to that day, we used to have a very good working relationship. He often would tease me and call me Mrs. Thatcher (former Prime Minister of the UK), and that was fine with me. People could call me anything they wanted, but threatening to sue me for doing nothing was vicious.

Back to the day of my meeting with Sara, on that December afternoon, she carefully looked at the loose sheets in front of her for a few minutes. Then pushed them back towards me on the table we were sitting at. She went to her desk, called somebody, came back, sat at the table and said, "Don't worry. I just talked to a LBL lawyer and he said they could not sue you over this." Then, while deeply in thought, as if remembering her own life, her own struggles, she looked at me and said, "You know. They hate our courage." And continued by saying, "I understand why you want to fight this boy club system, but the only way to get ahead is to compromise. You have to learn how to compromise." I interrupted her, crying out, "What fight? All I am doing here is barking, just like a dog. They don't even give me a stage to fight on." In the midst of voicing my frustration, I heard her teasing voice mercilessly saying, "So, you want to be taken seriously." As if I had no idea how tricky it would be to get that far. To calm me down, she started to talk about her own struggles with this abusive system. Her decision some years ago to leave science altogether, and open a restaurant. Then, suddenly, as if tired of hearing her own voice, she changed the subject and asked me, "Why are you quitting your job; what is your next plan?"

All along during the meeting, I was waiting for this opportunity to arise. So enthusiastically I told her about an unorthodox project of mine. The one I had been already working on since last year, albeit just a few hours every night after getting home from work at LBL. It was basically a statistical analysis of tens of thousands of people. The subject of my interest was a major human attribute that I thought could be a complex genetic trait (see p. 91).

At first, she seemed genuinely intrigued by my scientific rationale for this seemingly unscientific subject matter. But later she laughed and said, "I am not a geneticist, but I think you are crazy." Meaning that I could never get any academic funding for such a crazy idea. Then she said, "I was in Europe last week for a conference, and coming back I read a book about Barbara McClintock; have you read this book? You remind me of her." I said, "Yes, I have." And in my head I thought, and yes, I know. Like her, I can live in my own world, undisturbed by others' rules on how to act, and how to think.

Barbara McClintock was the crazy old lady of genetics, who was awarded the Nobel Prize in 1983 at age 81, some 30 years after she finished her major works in defining the transposable elements in maize (corn). She worked like a scientific outcast, and never could hold on to any academic position. Therefore, early in her scientific career, she was banished to a scientific Siberia, a small agriculture station in Long Island. It happened that a few decades later, by the efforts of several early molecular biologists working on the structure of virus, that little station became the world-renowned Cold Spring Harbor Laboratory. Had it remained an agriculture station, probably the world would never have noticed her unique understanding of genetics principles. When Sara mentioned her name, I suddenly remembered that one of her papers was the very first paper that I was assigned to discuss in our weekly lab meeting, way back in the first year of my PhD program in evolutionary genetics. I reminded Sara of the irony of her speech at her Nobel ceremony. She said something to the effect that, "Even though you have awarded me this prize today, you really have not yet grasped what I am talking about." She was referring to the fact that she called her little transposing genomic DNA "controlling elements" and not simply "transposable elements," as they are commonly called now.

Before her, most geneticists assumed that genes were fixed entities arranged on chromosomes like little beads on a necklace. She believed that the involvement of the transposable elements is crucial in the early stages of organismal development. Whereas later works by other scientists focused mainly on their disrupting actions, knocking out the function of genes by transposing and inserting themselves into them randomly. I still find it very amusing how we scientists sometimes so easily are able to alter the direction of others' scientific findings, and to fit them right into our own particular contemporary worldviews.

Time to Exit

My meeting with Sara took about two hours. That was 13 years ago, and I never could subscribe to her advice of compromising, much as I wanted to. It never made sense to me to commit myself to the standards set by other people for me. The problem with that approach was that I always had much higher standards for myself than they had for me. Or maybe the problem was, as later one of my older supervisors at the biotech company had told me, that people were afraid of me.

He was a scientist, and also a 30-year veteran of high positions in government. Maybc by sheer experience of dealing with so many different people, he knew what he was talking about. Although I could never figure out what could be so scary about an always smiling, five foot tall, 110 lb foreign woman with a bad accent, in this land of giants. I suppose they hated my guts, my confidence. As a matter of fact, many times I did not like my confidence that much, either. It was always such a hurdle for me when dealing with less secure people.

After leaving the biotech company in 2002, I had no desire to go to another company and work as an R&D scientist again. I figured out I would end up with the same people, only this time they would have different names. In the current corporate culture, all workers are bound to fill a few well-defined niches, with no room for individuality. I was rather interested in finding a job in bioinformatics, the newest field in genetics, and a very promising and exciting endeavor. In the early years of the publicly funded Human Genome Project, it became evident that new mathematical tools and innovative software programs were needed to handle the avalanche of data pouring from high throughput DNA sequencing machines.

These tools would be used for characterizing and categorizing DNA sequences (and proteins) across human genome. These tools could also be used for comparing DNA and protein sequences between different organisms, including humans, to better understand evolutionary processes shaping our genomes. Furthermore, the data could be used, in the growing conservative America, to show the extent of similarity between our DNA and those of our distant relatives, the great apes. So, at odds with the teachings of most religious books, it would be an indication of our shared heritage with the great apes and other organisms.

My interest in bioinformatics also seemed to be in the same direction of working directly with DNA for over 20 years–more like a natural progression of understanding DNA at its different levels. In my graduate work and my first postdoc position, I did extensive research on the evolutionary aspects of the alcohol dehydrogenase gene and its regulatory elements (see p. 130), as well as its enzyme system in *Drosophila*, which is the scientific name for fruit fly. This gene makes a protein with specific enzymatic properties, enabling this organism to survive in environments with high alcohol concentration like rotting fruits. It is, arguably, the most studied gene and enzyme system in genetics.

Afterwards, unlike many other geneticists who usually stick to their original gene of interest, I ended up working with some general aspects of DNA itself. At UCLA, I studied DNA-carcinogen adducts in mice, that is, how several cancer causing airborne chemicals bind to DNA and alter its structure and function. At UCSF, I studied the interactions between DNA and Bleomycin, an old and powerful cancer drug to better understand how specific drug molecules are able to target DNA, already affected by cancer causing agents. At the LBL (Lawrence Berkeley Lab) Genome Center, I was involved in different methods of mapping human chromosome 21, and I also worked with various human genomic libraries to isolate DNA probes associated with various human diseases. Later, at the biotech company, as mentioned above, I worked on the development of a fully automated and high throughput capillary DNA sequencer. I also worked on the characterization of DNA forensic kits for human identification applications, such as forensic and paternity cases; and genotyped the DNA of several thousand humans sampled from a number of ethnic groups.

To further advance my goal of persuading a career in bioinformatics, I took a one-week workshop in bioinformatics offered by LBL in November 2002. After eight years, it was very nostalgic for me to be back at the LBL ground, high above the UC Berkeley campus; and see again the familiar faces of the staff of LBL's Computer Science Department, who used to work with us in the exciting early years of The Human Genome Project. There were about thirty attendees at the workshop, many of them with degrees in computer science, including one sitting next to me, who was a young Chinese-American guy with a PhD degree in computer science, just recently laid off by a startup company. Additionally, one of our course instructors was a young assistant professor in mathematics from UC Berkeley.

Although bioinformatics is rooted in genetics, it seemed obvious to me that it is going to be advanced mostly by young computer scientists and mathematicians willing to learn some basics of genetics. On the other side of the coin was me, a middle-aged geneticist willing to learn their crafts. Taking the course and learning about bioinformatics tools was the easy first step. The next step was finding an introductory level job and learning how to apply and further develop those tools. I was really enthusiastic about applying bioinformatics tools in the study of human olfactory (related to sense of smell) receptor pseudogenes. These are several hundred slightly mutated duplicate genes with no apparent expression or function, scattered all over the human genome.

Despite my enthusiasm, in my heart of hearts, I knew very well that I was not willing to start all over again in a new field. I did not want to go again through the hell of being managed by at least two levels of supervisors, this time definitely by much younger ones. My ego would not allow it. That was my limitation. At my age, and with my education and experience, I thought, I deserved to be my own boss.

Unfortunately, science is a very expensive enterprise. One can work in a print shop for some 10 years, for example, and then go and open one's own shop. But no scientist can go and open her/his lab to do pure research without some external funding. But finding external funding for unorthodox research projects, with no university affiliation, is almost impossible. I am talking from my own experience. Several years ago, I tried very hard to get some funding, but got nowhere.

In the late afternoon of the last day of our workshop, we gathered in the spacious LBL cafeteria, with its high ceiling, surrounded by glass walls, overlooking the majestic cypress trees scattered all around the campus, with San Francisco Bay fading in distance. After the last exchange of emails, we said our good byes. I decided not to take the short cut, and started walking down the long stairway connecting the cafeteria to a lower level, where we could catch the LBL shuttle bus. The air was kind of chilly, and the natural surroundings around me seemed all merged into a peaceful blue haze.

I was quite pleased for having the opportunity to take this bioinformatics workshop. I especially liked learning a new tool for comparative genomic analysis. Applying this tool concurrently to the DNA sequences of several organisms, including humans, showed a great amount of genetic conservation among different organisms (see p.138). Of course, we were used to seeing this level of conservation for important house keeping genes. But now, using this tool, we could see vast expansions of genomic deserts, the areas with no apparent genes, also being greatly conserved among different organisms. Based on Darwinian selection, as these areas have no apparent function, they should not be under selective pressure, and hence they should be riddled with random mutations and show very little sequence conservation among different organisms. What an interesting result!

Soon more and more secrets of our evolutionary past will be unraveled. There are going to be lots of surprises for conservative geneticists. There are going to be blizzards of DNA data, some contradicting each other, and amidst such chaos, a new way of evolutionary thinking will emerge. What an exciting future awaits the genetics research.

I got off the LBL shuttle bus at the Berkeley BART station. It started raining very hard and it was getting very windy. I rushed into the station and went down to catch the San Francisco train. I settled into a seat next to the window with an unsettled mood. Soon, while the train was zooming ultra fast in the darkness of the under water tunnel, I started to think about my own uncertain future.

At age 49, with no retirement plan, and a high house mortgage, I had in front of me the tremendous task of inventing for myself a way to be my own boss, while still carrying with me the heavy burden of a scientific past. I thought, no matter what, I will still have my *Science* magazine and *Genetics* journal. The UCSF modern and gracious library with hundreds of scientific journals is just 15 minutes away, and the Internet is always waiting to search for me whatever comes to my mind.

I will never leave genetics behind. Great times in genetics are ahead of us.

Blessing by Knowledge, Another Ancient Heritage

So Said Zoroaster

From the very beginning, the Chinese community of San Francisco did not like the location of the new Asian Art Museum in the Civic Center. They said it had ghosts. The new museum, housed in the former San Francisco Public Library building, was reopened in 2003 after the library was moved to a new modern building across the street a few years ago. The last time I saw the old and stately building, before its extensive renovation and expansion, it definitely reminded me of a majestic ghost house. It had a grand marble central staircase, surrounded with dark narrow corridors, and small Gothic style windows, high above the floor. And it was ghostly quiet.

I did not know it then, but later I learned that, at the turn of century, this area of San Francisco was indeed an old cemetery. Several decades later, however, all cemeteries in San Francisco, including this one, were moved to Colma, an exclusive cemetery town south of San Francisco. Still, during the recent building expansion of the new museum, a few skeletons were resurfaced, and unnerved the Chinese community. A couple of years ago, when I walked for the first time inside the new museum, there was no more ghostly feeling, only grand spaces filled with natural sunlight. But when I left the building a few hours later, I felt as uneasy as the Chinese community, though for a completely different reason. As expected, there were big collections of Chinese and Japanese art works, in addition to some Korean ones. Large spaces were also devoted to India and several Southeast Asian countries. Being an Iranian, I was really excited to see some old artworks from Iran, the mighty Persian Empire of 2,500 to 1,400 years ago; instead all I saw were a few ceramic pots and a handful of very small and elaborate old silver coins. I thought, is that all? It is not fair.

It is true that, at the present time, Iran, a vast country with about 70 million people, is no longer a big player on the world stage. Nevertheless, one would expect that, at least in a museum setting, it deserved more attention. Later, I learned that it was not just me who was thinking Iran's past contributions to world civilization are being neglected. In 2005, the British Museum arranged a four-month exhibition in London called, "Forgotten Empire, the World of Ancient Persia." In fact, currently, Iran and other Middle Eastern countries are not considered part of Asia anymore. The Middle East, to outsiders, has become its own continent or rather a big Island floating on vast oil reserves and occupied by hairy looking people, worshiping a strange god, seemingly not deserving the trust of a world community fast growing very comfortable with standards and stereotypes.

Originally, I regarded this notion that the Middle East has become its own continent as a sentimental view, until I got into a bizarre argument with a Japanese friend. Actually, I was quite surprised when, during a conversation a few months ago, she insisted that the countries in the Middle East are not part of Asia. I had to remind her that every country belongs to a continent. But she was not satisfied, and insisted again that the Middle East is the Middle East. I reminded her of the Asian soccer games, and the fact that Iran and other Middle Eastern countries always participate in these games. Still, she was not sure, until I showed her a page from the *New York Times Almanac*, stating that Iran was in western Asia. Then she stared at me in disbelief and said nothing. What an annoying surprise it must have been to her.

I could have told her, how about considering Japan as another Pacific island and not part of Asia (after all, that is exactly what, geographically speaking, Japan is)? But we both knew very well that our discussion was not about the geography of Asia. It was about its people. But I said nothing more. Knowing that she was a very kind person, I thought it was not her fault that she had such a biased view. Sometimes our cultural differences lead us into believing that we are above other people, most often according to some distorted standards.

On another occasion, she surprised me again after I showed her a family picture of mine, and saw her sudden distrustful reaction. The picture showed my newly wed niece and her husband on their wedding day that was taken place recently in Tehran. After looking at the picture, she promptly said, "Her husband looks scary. Tell him to shave off his beard." Ironically, to me, he looked just like a young handsome guy with a small patch of beard barely covering the lower part of his chin.

Well, no matter how outsiders look at Iranians and other Middle Eastern people, we regard ourselves as warm and open people, always eager to know other people and the world around us. I remember that the front door of our house in Saba (Shiba) Garden neighborhood of Tehran was always open during afternoon hours, when endless kids, friends of my five brothers and sisters, would come and go as they pleased.

One day a little shy five-year old boy came in looking for my youngest sister. We soon learned that his name was Salar, meaning "top army general," a very unfitting name for such a tiny boy. He was so cute that we all gathered around him and started asking about his whereabouts. Feeling at the center of our attention, he glanced at one of the schoolbooks scattered around the carpet, where he was sitting. Then, struggling to read a line on its cover, he finally uttered, "Becomes enabled who becomes learned." We all got very excited that this little one indeed could read.

This line was written on the cover of every textbook in Iran from first to twelfth grade, in small black font on the top or bottom left corner. It is the first part of a line of poetry by an Iranian epic poet called Ferdowsi, who wrote it some 1,000 years ago. Its second part declares, "Through knowledge the old heart becomes young." The concept came from our Zoroastrian heritage, long before we Iranians became a predominantly Muslim people.

In his famous book, *So Said Zoroaster*, Nietzsche, the 19th century German philosopher, blames Zoroaster for bringing morality to the human psyche. Concepts of good and evil as exemplified by Ahura Mazda and Ahriman (Supreme God and Satan), paradise and hell, guardian angels, the Judgment Day, the soul's immortality and other supernatural concepts, all were first introduced through his teachings and spread probably from Central Asia to the Middle East and later became known to the whole world through the books of the Old Testament of the Jewish people.

Zoroaster, or Zartosht as we call him in Farsi, was an Iranian prophet who lived between 3,000 to 6,000 years ago (see p. 237), and as we learned in school, through his book, Avesta, he preached a path of human salvation based on "good thoughts, good words, good deeds" because according to him, only by doing good, we would be able to defeat evil in the eternal fight between good and evil. Avesta and other Zoroastrian books gradually disappeared from Iranian collective memory after

Arab armies brought Islam to Iran some fourteen centuries ago, and for some 300 years, Arabic, the language of the Quran, became our official language and replaced Farsi.

However, Iranian people stubbornly retained their traditional New Year and celebrated it on the first day of spring, as Zoroastrians did for thousands of years, according to their elaborate ancient ceremonies. Today, this tradition, called *Noruz* in Farsi, meaning "new day," is still celebrated in Iran as well as in several countries in Central Asia. It is, indeed, one of the oldest celebrations on earth.

Farsi, the official Iranian language and one of its 75 languages and dialects, came back to our tongues some 1,000 years ago through Ferdowsi's monumental epic poetry book called *Shahnameh*, a record of ancient Iranian kings and champions. He spent 30 years writing this poetry book, which he based on the stories gathered by many scholars before him, at the time of the Sassanid Empire of Iran. Apparently, by the order of government, they went village-to-village collecting information about Iran's past history.

The mythical characters in this book not only shed light on history, but also the prehistory of the people who lived in and around Central Asia as the original Indo-European people. Some of these people later migrated south to the Iranian plateau and the Indian subcontinent, mixing with the indigenous people of these regions.

Ferdowsi, in his book, proudly tells us that he has gathered stories going back 30,000 years. Many of his main Zoroastrian characters and concepts such as Jamshid, a mythical king (similar to Yima in Indian legends), *Jaam-e Jam*, by some accounts the original Holly Grail, and *farr*, an enigmatic Iranian supernatural concept (see p. 278), vaguely related to human intellect, in my opinion, surfaced again many centuries later in the poetry of other Iranian great poets, many of them Sufi Muslims such as Hafez and Rumi.

An outcome of the revival of the old Iranian national identity was the formation of a uniquely Iranian philosophy enhanced both by Zoroastrian and Islamic ideals. Thus, Ferdowsi fulfilled his own prophecy as stated in one of his poems that, "I will not ever die, as I am alive by sowing the seed of word." Not only did he spread Farsi words, he also initiated in Iran a poetic culture in which words rhymed together to entertain and please the heart as well as inform the mind.

Maybe the world, currently, prefers to think of Iranians as a bunch of terrorist-minded people. But we call ourselves people of the "land of flower and nightingale." By some accounts, during the last 1,000 years, Iranians have produced some 700 national poets. First, Ferdowsi taught us history and prehistory. Later, Rumi and Hafez elated us with secrets of Gnostic knowledge, as will be discussed in a later chapter. Indeed Ferdowsi has awakened in us our legendary storytelling nature, which has amused the world for centuries through the stories of *One Thousand and One Nights.*

These days, Western nations tend to regard people of the Middle East as backward people resisting any notion of progress, but about 1,000 years ago, around Ferdowsi's time, we were among the front runners of a new scientific era. The quest for experimental knowledge, as we know it now, was started by alchemists in the Middle Ages. One of the chief aims of alchemy, an early form of chemistry, was to change base metals into gold and silver. Even though Zacharia-e Razi (Rhazes in Latin, 865 to 925 AD), a 9th century Iranian alchemist, could not find the much sought after elixir to make gold, he went on to discover alcohol in the process. Later as a physician, he laid the foundation for modern clinical medicine. Even a few centuries before him, at the time of the Sassanid Empire of Iran, there was an institute in the city of Gondishapour (presently in southern Iran), where human dissection was taught to medical students.

The book *Al Qanun of Medicine*, also called *Canon of Medicine*, serving as the standard medical textbook in the European universities for seven centuries, up until the 18th century, was written by the great Iranian physician, scientist and philosopher, Ibn-e Sina (known as Avicenna in the West, 980 to 1037AD), who was of Tajik descent (see p. 114). Although he lived some six centuries before the European Renaissance, even the West now acknowledges his contribution to the Renaissance, by advocating the use of reasoning and logic as a way to gain knowledge. He is regarded, in the West, as the father of modern medicine particularly due to his role in the introduction of experimentation in studying physiology.

Abu Ali Sina, as we call him in Iran, apparently had an amazing memory. The legend has it that he could remember his childhood as far as being just a few months old. According to a story, his first memory of the sky was a meshed blue image. The reason it seemed like a meshed surface to him was because, as an infant, his mother, while working outside, had the habit of leaving him laid down under a sieve to keep flies away.

Another fascinating tale we know about him is how in one case he could find the house of a person he was looking for simply by checking the pulse of one of his patients. As the story goes, one day he was summoned to the house of a patient burning with a very high fever. Right away, he knew that it must be a case of a teenage girl being in love. But nobody knew about the whereabouts of the guy she was supposedly in love with. She herself was too sick or too shy to talk about it. So by keeping his finger on her pulse, he started to call the name of each street in their town, one by one, until he mentioned one street that made her pulse beat slightly differently. He continued applying his unusual method until he found out about a particular house in that street. Then, he ordered someone to go there and bring the young guy to see her. Surely enough, he was right.

Mixing Faith With Wisdom

As children, as we learned the virtues of gaining knowledge through reasoning at school, in the streets we were constantly exposed to another type of learning, learning through faith. Just like the composite nature of the culture I was raised in, I have two distinct names. Not one first name and one middle name, as is common here in the West, but like many Iranians of my age, I was given two first names.

One is my officially registered name, Farideh, meaning something like "uniquely created," an Iranian name given by my aunt from my mother's side. All my school friends and associates, and the new people I have ever met, know me only by that name. My other name is Mansooreh, meaning "victorious," an Arabic name, from the language of the Quran. It was given by my grandmother from my father's side, and was recorded in the back page of her old Quran. Most of my family members know me only by that name.

Sometimes, I wonder if someday someone puts a pistol on back of my head, and asks for my real name, what would my answer be? But without any pistol, I just feel I am a person with two names and two interlocked identities. Interestingly, my older sister, about four years my senior, has only one name, an Arabic name. My youngest sibling, another sister about nine years younger than me, has also one name, an Iranian name. It may or may not represent a trend on how fast during just one decade, in the 1950s, we Iranians switched from an Islamic identity to an Iranian one. Or more likely, it just could show how fast my grandmother lost her powerful grip on her eldest son and her daughter-in-law, that is, my parents; and was not welcomed in naming my youngest sibling.

In the early 1960s, when I was a little kid growing up in the upscale northern part of Tehran, there were not many practicing Muslims around us, except our grandmothers. However, we definitely experienced a religious culture, a fervent Shia culture for that matter, represented commonly by masterfully painted images of Shiite imams hanged on the walls of many downtown shops, and tiny local shrines associated with numerous *imamzadeh*. Generally regarded as saints, they were the offspring of eleven Shia imams, and were most probably a genetic mixture of Iranians and Arabs.

I remember the day before I came to the US, some 30 years ago, my mom insisted on taking me to Imamzadeh Saleh Shrine, in the Tajrish area, north of Tehran. And as if it was her last wish, I complied. I was 22 and she was just 44 years old. She must have been so worried to see me leaving her nest that she would do anything to calm herself down. That was before the age of Prozac and other powerful mood-changing drugs.

Recently, when I checked the Internet for a picture of the inside of this well-known shrine in Tehran, it was not at all like the picture I had kept in my head for so many years since I left Tehran some 30 years ago. The picture shown on the Internet was typical of the Shiite's grand shrines with an elaborate large net-like metal enclosure, either in pure gold or silver, surrounding the grave of a saint, in a room illuminated with several majestic chandeliers. But the picture I had in my mind was a little room with no windows, all walls draped in green, the color of Islam. In the middle of the room, there was a raised platform, supposedly the tomb of Imamzadeh Saleh, also covered by a thick green velvet fabric. As I remember, everywhere I looked, in that room, it was green and indeed a very memorable sight.

So, I wondered, what happened? The only thing I could think of was either that after Islamic Revolution of Iran in 1979, the shrine was upgraded, or maybe I had completely forgotten its inside and replaced it with the image of a typical little local shrine that I remember seeing many times in my childhood. As children, the most anticipated yearly religious event for us was the two nights of *Ashura* and *Tasua* –elaborate street ceremonies sponsored by rich merchants of Bazaar, a huge and labyrinthine mosque-like structure in central Tehran, going back maybe two centuries in time. During those nights, it was very exiting to be with other people, many dressed in black, marching in the local streets chanting and learning about the cruel death of Imam Hussein and his brother Abolfazl-e Abass during the Karbala tragedy, some 1,300 years ago (Karbala is a city in central Iraq).

Abolfazl-e Abass was my mother's personal saint, as he was to so many other Iranian women who showed great devotion to him. Whether his personification in the battlefield of Karbala was real, or seamless fabrication, we will never know. After revisiting his legend last year on the Internet, some 45 years after my childhood memories, my tears started to fall down slowly, though I am not typically a crying type of person. I right away thought that the West needs to come up with much stronger storylines than *Star Wars* and *Lord of the Rings* to lure away the kids in the Muslim world.

The legacy of Abolfazl is not only because of his battlefield bravery. His name means "lord of virtues" and he is referred to as *ghamar-e bani Hashem* which means the "moon of the clan of Hashem" (in Eastern cultures "moon" is considered as the most beautiful object in nature, and as its brightness is from reflected light, it is viewed as a mystical embodiment of divine inspiration). He was tall, handsome, full of virtues, and a son of Imam Ali (the first Shiite imam, referred to as "prince of believers"). But he was still regarded as an underdog, which is the key point of his legacy. He was the underdog to his much older brother Imam Hussein (the third Shiite imam, referred to as "master of martyrs") because unlike him, he didn't have a perfect lineage.

Imam Hussein's mother was Fatemeh-e Zahra, the beloved daughter of the Prophet Mohammad; and his father was Imam

Ali, who was also a cousin of the prophet. Whereas Abolfazl's mother was another wife of Imam Ali, whom he married later after his wife, Fatemeh-e Zahra, had died at the tender age of eighteen. So, Abolfazl from an early age, under his mother's guidance, dedicated himself to serving his older brother and his brother's children. He lost his life to a cruel death, knowingly going through a barrage of enemy arrows, hoping against all odds to bring water to his very thirsty, four-year old beloved niece, in a scorching day in that stormy desert of Karbala.

Centuries after his death, he is still regarded as a miracle worker and many people claim to see him in their vicinity every day; in just the same way, many Catholics believe in seeing their saints. Indeed, there are great similarities between Shia Muslims and Christian Catholics. Shias, like Catholics, because of their devotion to their saints, are routinely mocked by their mainstream brethrens. But as Italian Catholics became master painters of the Renaissance, Shia Muslims, a small fraction of the 1.3 billion Muslims worldwide, became master designers and builders of the Islamic world.

In his book, *Islamic Art and Architecture, from Isfahan to the Taj Mahal* (2002, Thames & Hudson), Henri Stierlin, a renowned architectural historian, argues fervently for the spread of "Persian style" architecture throughout the Middle East, and as far as the Indian subcontinent and Central Asia between the 9th and 19th centuries. The Taj Mahal, in Agra, India, one of the most majestic and enchanting architectural masterpieces in the world, is said to have been designed by an Iranian architect named Ustad Isa by the order of Shah Jahan, the Mughal Muslim emperor of India. He built the Taj Mahal as a tomb in memory of his wife Momtaz Mahal, an Iranian woman whom he married in 1612. She died at age 38 while delivering their fourteenth child.

Since painting the human image was traditionally forbidden in Islam, much of the artistic creativity of Muslims went into calligraphy, arabesque and other ceramic designs to adorn mosques, palaces, bazaars, tombs, bathhouses and other public buildings. The tiling designs of many of these structures are, mathematically speaking, amazingly intricate. Indeed, there was recently an article in *Science* magazine (23 February 2007, volume 315, p.1106), suggesting that the unknown architects of some of the Islamic medieval structures, made between the 13th and 15th centuries, must have discovered a piece of mathematics so sophisticated that was defined in the West only about 30 years ago (called quasi-crystalline Penrose patterns).

Isfahan, a city in central Iran and the capital of the Safavid Empire of Iran, was designed and built about 500 years ago and soon became the most populated city of its time in the world. Some years ago, while watching a TV program on PBS named, *Islam, the Empire of Faith*, I heard a Turkish historian calling Isfahan the most magnificent city on earth because of its exquisite beauty. I found her statement very refreshing considering that her own city, Istanbul, is widely known as the crown jewel of Islamic architecture.

A rather small mosque, Masjed-e Shaikh Lotfollah, in Isfahan's famous grand plaza, is still my favorite masterpiece. I saw that building when I was 16 years old, and after seeing hundreds of masterpieces in Europe and the Americas since then, to me that is still the most secluded and peaceful place on earth. Entering it from a small opening, it suddenly rises up to a huge blue ceramic dome, surrounded by massive walls of blue and yellow ceramics. Amid this grand design, somehow it still manages to feel very quiet and tranquil. There is a place in Northern California that gives me the same kind of peace as if I am going to meet God. While going down an endless flight of narrow steps to get to the sea level to the lighthouse at Point Rays National Park in Marin County, north of San Francisco, again suddenly one feels completely surrounded by a haze of blue color, this time, the calm, majestic Pacific Ocean.

Islam conquered Iran some 1,400 years ago, after several years of intense fighting. As we Iranians always like to say, a bunch of nomadic tribes from the Arabian deserts brought down the mighty Sassanid Empire of Iran, a super power of its time. But the fact is that Islam conquered Iran not only by the sword, but also with the promise of equality among all humans, regardless of tribal origin. Iranians supposedly opened the gates of their cities themselves to the invading Arab forces, bringing Islam to Iran because they were apparently fed up with their own rulers.

By the time Islam reached Iranian borders, the Sassanid Empire had already lost its unity through palace internal fighting between various factions, triggered by powerful Zoroastrian high priests called *mubed-e mubedan*. A vast nation of different people was already demoralized through the succession of six kings during a period of just seven years.

Another factor in the Iranian defeat was the meddling by Sassanid's main rival at the time, the Roman Empire, which, by supplying the Arab army, helped put more pressure on an already tattered government. Ironically, Christianity itself indirectly helped establish Islam around the world. After the Iranian gates of the Sassanid Empire fell down to the Islamic army, the spread of Islam from Spain to Indonesia was just a matter of time. But Islam itself did not survive intact in Iran. Five hundred years ago, Shia became Iran's official state religion, at the time of the Safavid Empire. Basically it was a political gesture by Safavid rulers to ward off the Sunni Muslim Turks of the neighboring Ottoman Empire.

Shia Islam, by its main emphasis on *imamat*, the succession of 12 imams starting with Imam Ali, the younger cousin of the Prophet Mohammad, and all directly from one line, the line of Fatemeh-e Zahra, the beloved daughter of the prophet, in essence brought back the Iranian enigmatic and ancient fascination with the concept of the divine representation of kings.

Many of these imams, not political rulers in any sense, were great Islamic thinkers. Their scholarly writings, propagated through the Internet these days, as never before, are testimonials to their sincere devotion to Islamic concepts. Iranians' great devotion to *imamat* and their imams was also enhanced by the fact that Shahrbanoo, the Zoroastrian daughter of the last king of the Sassanid Empire, later supposedly became a wife of Imam Hussein, the third Shia imam, and therefore, all imams afterwards descended from her line.

A mountain south of Tehran, named after her, Bibi Shahrbanoo, still reminds the city travelers of the time when we mixed with our enemy and became an even more diverse nation. Mixing with Arabs not only helped sustain Islam in Iran, but also brought Zoroastrianism to Islam. The great Islamic philosophers, such as Sohrawardi, mixed Zoroastrian concepts with traditional Islam (see p. 295), and at the same time helped establish a martyrdom cult by going against mainstream Islam.

Despite the fact that the West tends to ignore any contribution by women or respect for women in Islamic cultures, the *imamat* in Shia is defined mainly by the contribution of two women–one a daughter of a prophet, Fatemeh-e Zahra; and the other a daughter of a king, Shahrbanoo. *Imamat* indeed was shaped with their genetic contributions.

Another Muslim woman, Seyedeh Zainab, a towering figure in Shia Islam and the elder sister of the third imam, on the other hand, defines *imamat* in a spiritual sense. If there were not a courageous Zainab, there would be no memory of the Karbala tragedy left in people's minds, and for that matter there would be no Shia Islam.

After witnessing the slaughter of all her male family members in Karbala (except one, her nephew, the fourth imam), one by one, at the hand of a much powerful rival Islamic army, Zainab gathered the remaining women and children and went home. But instead of fading into history by getting scared and keeping quiet, she made it her business to go after the ruling Muslim leaders to remind the public, over and over, of what they did to the prophet's family, all for the sake of power. One wonders, with the contribution of powerhouse Muslim women like Zainab, how current Islamic governments could justify not allowing equality in human rights for men and women alike.

Despite these inequalities, a large number of women in Islamic societies are presently pursuing higher education like never before. Two years ago, there was a cover story about science in Iran in *Science* magazine (16 September 2005, volume 309). It stated that currently about 20% of scientists working in Iranian universities are women, a figure twice higher than Japanese or Korean women. Not mentioned there, but according to other reports, over 50 percent of graduating medical students in Iranian universities are also women as well as over 65 percent of all university students.

Nevertheless, in a world becoming very comfortable with stereotypes and hastily moving forward towards a dull global culture, dominated by the exhausted Western ideals, it has become increasingly easier for the world community to ignore all these modern-day achievements and focus only on the traditionalism of Islamic cultures.

Land of Many Traditions

Religious fervor among Iranians did not start by adoration of their Muslim imams. Reading about Zoroastrian *mubed-e mubedan* (high priest) during the Sassanid Empire is like reading about Shiite ayatollahs. Indeed the founder of the Sassanid Empire came to power 1,700 years ago with the promise of bringing religious values back to the Iranian society. It turned out that it was not just about bringing back religious values, but about intolerance to other views as well. Two well-known Iranian prophets with other religious views appeared during this era. Both movements were suppressed and their leaders martyred with astonishing cruelty.

The first one was the Prophet Mani (see p. 254), who introduced a new religion combining Zoroastrianism, Christianity and Buddhism. His religion did not flourish in Iran but had widespread following in the West and Central Asia for centuries to come. His movement, known as Manicheans in the West, greatly influenced early Christianity. St. Augustine, one of the main founders of Christianity, was originally a Manichean before converting to Christianity (see p. 256). The second prophet was Mazdak, who advocated socialism through his religious teachings. His movement did not survive in Iran either.

Even now, we Iranians, like some other traditional Eastern cultures, do not show much tolerance towards other people's views and it goes at every level of our society. But a long time ago, Iran was the center of tolerance in a vast area of Asia known as the Persian Empire. Cyrus the Great, a Zoroastrian king of the Achaemenid Empire of Iran, who conquered Babylon in 539 BC, declared free religion for all 41 nations making up his empire. Later, during the reign of his successors, the Achaemenid Empire became the largest empire on the face of the earth.

A small baked-clay barrel, now commonly called "Cyrus the Great cylinder" and housed at the British Museum, was originally placed in the foundation of the city wall of ancient Babylon (now in central Iraq). His words, scribed in Babylonian Cuneiform on this cylinder, bring Cyrus back to life and he could be heard again addressing the city's people very warmly in a long and engaging speech. My favorite part is where he says, "I announce that I will respect the traditions, customs and religions of the nations of my empire and never let any of my governors and subordinates look down or insult them until I am alive." A replica of this cylinder, displayed at the United Nations, in New York City, represents one of the earliest charters of human rights. In 1971, its translation was published by the United Nations in all its official languages.

After the conquest of Babylon, he freed a large Jewish community there and allowed them to go back to Jerusalem and rebuild their temple. Cyrus is the most frequently named foreign figure in the Old Testament, and the only foreigner whom the Jewish people called Messiah, their savior. Some freed Jewish people supposedly chose to go to Iran with Cyrus and there they established one of the oldest Jewish communities in the world. The Jewish reverence for Cyrus was not merely literal and has survived over two and half millenniums among its people. The Iranian Jewish people still name their sons after Cyrus. Kurosh, his original name (Cyrus being the Greek name), is as common among Iranian Jewish people as it is among other Iranians.

Many years ago, while I was a postdoc at UCSF, I met two visiting organic chemists from Israel, Rachel and her husband Zeff, both orthodox Jews, originally from Poland. Right after we met, she informed me that in their daily religious practices, they follow many Jewish traditions borrowed from Iranian Jewish communities because their traditions represent the oldest traditions among all Jewish people.

I was raised in a new subdivision in the north of Tehran populated by Muslims as well as Jewish and Armenian Christians. I remember that once, when walking to a nearby store, a woman sitting on her doorstep, as it was common in those days, called me and said, "Little girl, come over here." Then she put me on her lap and asked me, "What is your name?" After hearing my name, she asked me, "Is your mom *Kalimi*?" After I said no, she let me go. *Kalimi* was the proper word for calling Jewish people and means a "special type of speaker," an "interlocutor," referring to Moses, who in Islam is called Mussa kalamullah, meaning Moses who brought the word of God. As children we were taught to call Jewish people *Kalimi* as a sign of respect for their long-held religious traditions.

As a religious minority in Iran, as in many other traditional societies, the Jewish people, as did other minorities, probably were not always as well treated as they would have liked to be. But since they were able to fully practice their ancient traditions, and also were financially prosperous, this points to the reality that historically the Muslim and Jewish people of the Middle East were always able to get along with each other, until the international politics of the 20th century created occupied lands there and infuriated the people of the Middle East.

Now, to establish again a more peaceful coexistence in this region and other regions, it is up to the United Nations to reinvent itself by granting true equality among all its member nations. Since all humans in essence are born, at the genetic level, equal to each other; their societies, cultures and belief systems are deemed to be regarded as variations on the same theme, and therefore, equal to each other as well. Without a doubt, this is going to be a daunting task for the world body to accomplish, especially with the current atmosphere in which the concepts of equality and "rule by the ruled"–in another word, democracy–have been deviously weakened by a capitalist machinery that through mesmerizing and attention-grabbing marketing strategies has brought "one brand only" mentality to the contemporary consumerist societies.

Going back to the Iranian Jewish community, by some accounts, in the '60s and '70s, on per capita terms, they may have been the richest community in the world. After the Islamic Revolution of Iran, many of them moved to the Los Angeles area, and currently, about one third of the real estate in downtown Los Angeles belongs to these people. Nevertheless, they are as homesick for Iran as any other Iranian expatriate. The Mizrahi Jewish people of Iran share a long history with other

Iranians, and to any outsider they look and behave very much the same.

A few years ago, in Washington DC, going from one airport to another, due to stormy weather and cancellation of some flights, I took a long taxi trip. The driver was also an Iranian, so we talked. He happened to be a Zoroastrian. While we were talking, as if there were something absolutely wrong with faith-based knowledge, he mentioned that whereas Islam is based on faith, Zoroastrianism is based on reasoning. That was the first time I became aware of such a profound demarcation between Iran's two prevailing religions, albeit at two different stages of its long history. Some Zoroastrians fled Iran at the time of Islamic invasion and settled in Bombay, India, establishing a thriving merchant community there called Parsi.

I remember, 30 years ago, when I had just started studying at Stanford University, one afternoon, at the entrance to my dormitory building in Escondido Village, an Indian woman, who later I learned was a postdoc in the Chemistry Department, dressed in a simple sari ran towards me, and with great excitement asked me whether I was a Parsi. I clearly remember that she looked kind of disappointed when I replied, "No." That was the first time I had heard of the Parsi community in Bombay.

There are now some 200,000 Zoroastrians worldwide, some still in Iran. Zubin Mehta, the famed Indian music conductor, a Parsi from Bombay, occasionally refers to his centuries old Iranian heritage. Indeed, with his Iranian name, Zubin, and his facial features, he looks as Iranian to me as any other Iranian. It seems his genetic heritage has not been diluted much during the past thirteen centuries. Zoroastrians, like the Jewish people, and many other ancient communities pressured by time and small population size, strongly defy inter-religion unions.

Recently I heard from my sister, who lives in Iran, a heart-breaking story about one of her Zoroastrian friends, who married a Muslim guy years ago. Regrettably, last year she was left out of a hefty inheritance because her mother, before dying, left all their family assets to her younger brother, who had married a Zoroastrian. When I heard about this incident, I remembered the Iranian taxi driver in Washington DC, who talked about Zoroastrianism being based on reasoning. And I wondered how could reasoning be applied to human complex situations, such as her mother's last action. Sometimes all we need is compassion, but sadly, regardless of our tribal origin or religion beliefs, as Hafez, the 14th century Iranian Sufi poet, observes in one of his poems, "If we look wisely, we are all hypocrites."

PART TWO

UNDOING A MINDSET

Apparently, as a byproduct of more recent technological advances, the human mind is poised to acquire a robotic imprint, enabling it to be more focused and productive on one hand, and narrow-minded on the other. However, this provocative development is nothing new to scientists, who routinely go through indoctrination in their particular fields of inquiry to be ordained with a very focused and purposeful mind. As an evolutionary geneticist, I have been there and know all about it.

Then comes a time, a desire or a need to reclaim one's mind, the once curious and improvising mind, leading to a process akin to reformatting a scientific mind into an independent thinker. But it is never easy to leave behind a well-trained scientific mind. Allowing oneself to be distracted is an essential first step in undertaking this risky journey. Some of the distractions that managed to unleash my systematic, scientific mind are underlined in the second part of this book.

A Room Full of Skulls

It was a typical foggy afternoon in San Francisco, and as usual I wished it was a bright sunny day. I climbed up the steps and entered the California Academy of Science building in Golden Gate Park. Inside it was packed with people, mostly middle-aged parents with their small children. The vast Academy building was scheduled for renovation and was going to close in three days. So its doors were opened to the public free of charge during these last days. I originally went there to see the show at the planetarium, but the line was so long that I decided to come back and see the show at a later session.

Meanwhile, I started to wander around the long corridors inside the building. Soon I found myself in a huge room full of skulls. Just like a kid in a toy store, I could not believe my eyes. I walked to the center of the room, stood there for a few seconds and then turned around and around. I was surrounded by three walls covered with skulls from top to bottom, and my mind was wondering about how come I never heard or read anything in the newspaper about such a magnificent exhibition.

The floor was covered with long tables with glass cases sitting on top of them, each filled with more skulls. On display, there was the skull of every vertebrate one could imagine. All skulls were more or less the same color, ivory-yellowish, seemingly newly minted and very clean, sitting in rows after rows in the glass cases, and each glass case was representing a different category of animal.

Just to have the chance to compare the skulls of so many creatures in the same room seemed so surreal to me. Each one was very unique and some were very bizarre, but almost all of them shared two characteristics: long and big jaws, and small and thick craniums, the skullcaps. Looking at these massive jaws with big teeth, one could see that living is all about eating and chewing. Having a big brain and thinking seemed to be almost an afterthought for evolutionary forces.

I looked and looked until I came across a glass case filled with the skulls of higher primates. Next to the massive skulls of the chimpanzee, orangutan and gorilla, I noticed a few delicate skulls, with small jaws and big fragile craniums, who were gazing at me with hollowed eye sockets. They looked so out of place that all I could think was, "What are you doing here, you poor little things, among all these big jaws?"

While a graduate student at Stanford, many years ago, I had taken several courses in paleoanthropology/human evolution. However, the two dimensional black and white pictures I had seen in text books never could reveal such a sharp distinction between the fragility of the human skull and the skulls of other higher primates. I stayed several more hours in that room, comparing different skulls over and over, and visited them again the next day. When, at last, I left that room and walked into the darkness of night to my parked car, I realized that I was not alone and had taken with me from that room some previously unthinkable thoughts.

To my utter disbelief, I had started to think for the first time as an evolutionary geneticist that there might be more to human evolution than was suggested by Darwin some 145 years ago.

What a disturbing thought. My poor scientific mind, after going through so many years of scientific training in several top-notched universities, now so easily had succumbed to some ugly unwanted ideas, just by spending a few hours in a room full of skulls.

With a thick fog settling in, I drove slowly through the dark and deserted back roads of the park, with a head that was becoming as foggy as the outside weather. I could vaguely see yet another uncharted territory opening up in front of me, inviting my beguiling mind to yet another enigmatic journey. This time I was not enchanted at all. I was not ready yet. I thought that I had learned my lesson.

My first scientific adventure into another uncharted territory spellbound me for more than two decades and ended nowhere (see p. 92), leaving me tumbling behind an otherwise promising scientific career. Still I do not regret it. It gave me a deep genetic insight, and fulfilled my dream of scientifically addressing an enigmatic subject that has managed to attract people for over 5,000 years; too bad for the genetic establishment that was not ready for it.

By the time I reached home, after going up and down the hilly streets of Monterey Heights, I still had those unwanted thoughts clinging to my head. I wished I could scatter them into the night fog to get lost and leave me alone. While unlocking the front door and entering the dark hallway, I remembered a well-known Iranian saying, "Whatever happens, it will be welcomed," and I reminded myself silently, be yourself again, be a brave heart.

My First Close Encounter with Quran-e Majid

I took the large bag to the kitchen, put it on the kitchen table and started to empty its contents. My brother was visiting me from LA and my mother had sent with him some Iranian food for me. I took out the frozen green mix, already chopped and sautéed in oil and portioned in several zip-lock plastic bags; and I placed them neatly in freezer. That was to make my favorite Iranian stew, *khorosh-e karafs*, which is a celery dish. Then, I started to nag about the other stuff that she had sent, like French feta cheese, and a kind of Iranian bread called *barbary*. I said, "She knows that I can find these items here in San Francisco, so why did she make your bag heavier by sending this stuff?" Looking a little puzzled upon hearing my stupid complaints, my brother said, "That's okay, you know how much she enjoys spoiling us."

While lifting the bag to put it aside, I noticed that it was still kind of heavy. So I looked inside and found something like a book placed in an especially made fabric case. I took it out and pierced inside its case by turning the fabric flap. The book had a beautiful sky blue soft cover and over it in big black font was written the word Quran in Farsi. Puzzled, I asked my brother, "Since when mom thinks that I should be reading the Quran?" He quickly replied, "No, that's not for you." Soon, I learned that my mom had sent it for my younger sister, a dentist, who also lives in San Francisco.

Then he added that an Iranian radio station in LA had recently talked a lot about this Quran, which is the first standalone Farsi translation of the Quran. In Iran, we always used to see the Quran in its original Arabic language, with its Farsi translation in smaller font written below each Arabic line. Apparently, my mom thought that the Farsi version might be of interest to my sister who is also an accomplished artist.

I took the book, still in its case, and carefully placed it on the leather covered round end table in the living room to give it to my sister later when she visited me. Several days passed and the blue book was still sitting on the end table untouched. It was very uncharacteristic of me to leave a book alone for several days. I am one of those people who cannot resist any written page. But this time I thought, what could I gain by reading a prayer book that only my grandmothers used to read? I already knew by heart two small *sureh* (section) from the Quran in Arabic, like most other Iranians do, and I thought that was enough for me in case I needed to pray.

A week later, I went to the end table to take something from its built-in cabinet and I noticed the blue book again. I casually lifted the fabric flap. Its cover looked so peaceful. It depicted a blue sky with fluffy white clouds, my favorite pieces of nature. It definitely did not look like the Qurans I remembered from Iran. They were all old looking Arabic books with dark hard covers.

Hesitantly, I took the book out of its case and placed it on the kitchen table to browse through it. As soon as I reached the book to open it, I was taken aback. I thought, maybe at first I should go and wash my hands. As a non-practicing Muslim, I felt very uneasy around this blue book. It looked like a typical book written in Farsi. But it was still the holy Quran, the book that we used to swear by constantly while kids in elementary school. Everyday, at least a few times, we would say, "To Quran-e Majid, I did not do that." Or "To Quran-e Majid, I did not say that." Then, I did not know what *majid* means, but now I know it means "noble" and "glorious."

When I was growing up in Iran before the Islamic Revolution, although neither my generation nor my parents' generation had read the Quran, we had great veneration for its physical presence. The way to handle it was to lift it up with both hands, bring it close to one's face, kiss it from the side closest to the face and then lift it a little higher and press it to the forehead. So now, like a little kid, I did not know how I was supposed to handle this blue book. If I kissed it, then how could it be like any other book?

Not sure what to do with this blue book, I just opened it to a page in the middle and gazed at the open pages in front of me. My eyes reached a sentence in the middle of the page, and I started to read it. It said something like, "After you die, on Judgment Day, we gather your remaining bones and from them, we bring you back to life." I thought, very interesting, it sounds like genetic cloning from a piece of bone. Then randomly I opened another page and it said something like, "You think God has sons and you have daughters?" Now I was really intrigued to know what was going on. I continued to read the next sentence, but the subject entirely changed. I reached for a pen to mark that sentence, but when I brought down the pen close to the page to mark it, my hand just stopped and I could not bring myself to mark the holy Quran. I thought, maybe I should mark that page with a piece of paper, but I could not do that either. I got so frustrated with myself for not being able to treat this blue book just like any other book that I closed it and put it back in its fabric case and left it on the end table again.

Then, as usual, I went downstairs and checked my mail. There it was, the latest issue of *Science* magazine. After browsing through it, I noticed an article about olfactory receptor genes, one of my favorite subjects. I sat at the kitchen table with the magazine in front of me. As I looked outside, two Monarch butterflies were dancing together around the outside doorknob, as they did every afternoon around this time. I could see the fog was moving in. I took a pen and started to read the science article and marking the sentences. Now I was happy.

The next morning, as soon as I opened my eyes, my mind began to drift to the enigmatic sentence I read in the Quran the day before about God's sons and humans' daughters. Although Christians always talk about Jesus as the Son of God, and we all know about the Greek gods with some human characteristics, I don't ever remember any mention of God having offspring in Islam. The God in my mind was always transcendent, absolutely apart from the material universe.

I tried to remember the exact sentence that I had read the day before, but I couldn't. I wished I had marked the sentence at that time. Then suddenly I thought that by reading the whole book, I could find that sentence again and any related topics. Suddenly, at last, there was a reason for me to read the Quran.

By the end of that year, about three years ago, I read the Quran not once, but three times. In my first reading, I found that particular sentence I was looking for, but not any related subject, and while reading, somewhere in the middle, I noticed an intriguing relationship between God and humans, more complicated than what I assumed from a few prayers that I knew by heart. That was the reason I read the Quran the second time. It seemed to me that God, although merciful to humans, was somehow very disappointed with humans, or rather with the human mind, its capacity for understanding and especially foreseeing the consequences of her/his actions. In numerous passages, it says, "God knows and you do not know." But

God's disappointment was not only because humans are inherently stupid; it was rather due to the fact that humans are arrogant as well. In several places, it said that humans do everything according to their wishes with no regard to God's universal order that all other creatures and planetary objects obediently follow. Another act of human arrogance, according to the Quran, was that since humans inherently lack the capacity to know by certainty, instead they would try to speculate all the time. Then, in one of the most enigmatic passages of the Quran, it says, "We offered some safekeeping load to the skies, earth, and mountains. They refused to accept it, but the human took charge of it because the human is cruel and ignorant/does not know." When reading this passage, I remembered that Hafez, the great Iranian Sufi poet of the 14th century, wrote a beautiful verse referring to the same enigmatic load. Then suddenly I realized what it could be all about. This book in essence was written as an attempt to contemplate more about that enigmatic verse.

The third time I read the Quran was to find passages about the creation of Adam, which grabbed my attention towards the end of my second reading. Adam, supposedly the forefather of all humans, according to the Quran, was created very uniquely, in the image of God. There is a beautiful passage in the Quran describing the event in which God asked Satan, originally an angel, and other angels, to prostrate in front of the newly minted Adam.

Later, as I was moving the Quran from the end table to the bookshelf in the living room, I thought again about the enigmatic load that ignorant humans accepted from God. Then, I lost my focus and stared outside through the window. I could see the surge of afternoon fog towards the window, then stopping there and greeting me as usual. It reminded me again that there was no enduring sunshine in the magical San Francisco, only endless variations of the playful fog.

Poetry of Hafez and His Gnostic Tongue

I pushed down the button on the CD player and Molana (Rumi), the Iranian Sufi poet, who is arguably the most read poet in the US, started to talk to me in Farsi after so many years:

I am the servant of the moon, except moon,
not say anything;
with me except the talk of honey and sugar,
not say anything;
not say any talk of suffering, except the talk of treasure,
not say anything;
from this unawareness, not suffer, not say anything...

In just two minutes, it left me completely drained, as if a monsoon rain had just washed away not only dirt but my flesh and bones as well, exposing my bare soul to infinite emptiness. I felt somehow uncomfortable in his poetry universe. It was too liberating for me. I desperately needed my body back to rest my soul on it, to keep it anchored.

A few months ago, when my sister was coming from Iran to the US for her yearly visit, I asked her to bring me more audio readings of Hafez' poetry (several years ago, she had brought me a tape containing 20 of his poems). As usual, she was more than happy to accommodate my request. And right away she suggested bringing me not only a CD of Hafez, but those of Molana and Omar-e Khayyam as well. But I said, "No, please, right now I am concentrating on Hafez' poetry, and I am not in the mood for Molana's poetry."

Actually, as I recalled my high school years in Tehran, studying his poetry in our literature class, I never was in the mood for his exquisitely restless divine soul, like a rushing river of liberated words moving forward with the force of volcanic molten rocks. But my sister being herself, never minding what you wanted, but what she wanted to do for you, brought me not only a CD of Hafez' poetry, but also those of Molana and Khayyam.

One day, at last, I ventured out to get reacquainted with Molana's poetry as recited most clearly and vividly by Ahmad Shamloo, a contemporary modern Iranian poet, who died a few years ago, God bless his soul. Listening to Molana's poetry now, sometimes I give in, and take his divine invitation to free our neglected souls, with no resistance whatsoever. But when I am in a defiant mood, I tend to talk back, reciting a somewhat contradictory poetry piece by Hafez, who was born about a century after Molana in the 14th century:

How I circle around the sky of the holy realm,
since in the configuration blueprint,
I was nailed down to the body...

Hafez, the master of all masters of Persian poetry, after some 600 years, still lives in the hearts of endless Iranians, sharing the secrets he knows with our anxious contemporary souls. His poetry is revered so much that his book sits next to the Quran, on a ceremonial table every spring for 13 days during the Iranian New Year called *Noruz*, a traditional 5,000-year-old Zoroastrian ceremony. Ironically, the passage of time has replaced the original book of ancient Zoroastrian religion, Avesta, with these two newer books revered by so many Iranians.

Another symbolic gesture, indicating our sincere devotion to Hafez' poetry and philosophy, is a common practice that many modern Iranian brides undertake. The ones who refuse any future monetary compensation, in case of divorce, and routinely just ask for a volume of his poetry book in their written marriage contracts. He has earned our immense spiritual trust by painting himself in his poetry as an everyday human passing pensively through everyday life. But while enduring, he chooses to be daring and defiant. In one of his most celebrated poems, he talks about opening the ceiling of the universe and designing a new pattern devoid of any sorrow.

As a devout Muslim, who according to himself, as stated in one of his poems, could recite the Quran by heart in 14 versions, his desire to redesign the universe is a blasphemous thought. Who is he to go against God's will, one may ask? But the way he says it in just eight simple words is so powerful and grandiose, yet graceful, that turns his act of challenging divinity into his basic human right. Not feeling completely at ease with divinity, the way Rumi does, nevertheless, he is so much in awe of the universe, both its divine and physical aspects, that it seems as if he could act like a petitioner for us, echoing our earthy human pains to higher universal sources.

With an undisputed technical poetic mastery, still the magic of his poetry lies in the fact that he, like Rumi, is a revered mystique Sufi. During the last 600 years, some 600 books have been written about him, trying to figure out his poetry. Like most poets, he frequently talks about heaven and angels. But in his poetry, these are not merely symbols of spirituality. What fascinates me most about his poetry is that they seem like real places and real beings, very much involved in the early events of the human prehistoric life.

His poetry book is commonly used as a divination source by millions of Iranians everyday, when they feel under duress and seek the presence and advice of a caring companion. Customarily, before opening his book randomly for divination, he is called upon by saying, "You Master Hafez of Shiraz (his city), you are intimate to every secret…" And if he answers you by one of his poems like, "The lost Joseph will come back to Canaan, don't be sad….," for a moment, you will feel so blessed that you almost forget your sorrows and troublesome thoughts.

Arab poets called him *lasan-ol ghaib*, the one with the Gnostic tongue because it seems that the essence of his poetry springs from an inner revelatory fountain. The belief that an elite few, including Hafez, were in possession of *elm-e ghaib*, is a central tenet of Iranian philosophy as well as those of other people of the Middle East. *Elm* means "knowledge," so someone who has access to *elm-e ghaib* is said "to have the tongue of *ghaib"* or "Gnostic tongue." *Ghaib* refers to a hidden universe of secret knowledge veiled from ordinary people, and it is about a superior spiritual knowledge, which is supposedly accessible mainly through self-illumination, by means of living a disciplined and non-materialistic life.

The notion of *elm-e ghaib* is frequently expressed in everyday conversation among Iranians. It is very common to hear people saying sarcastically, "How do you know that? Do you have *elm-e ghaib*?" In English, *ghaib* has been commonly translated as an "invisible or a mysterious world." It has also been technically defined as the "super conscious" to differentiate it from the "subconscious." Some sages, who have witnessed both these mental activities, have described them as being black in color. However, they can recognize each distinctly, due to the colors that appear before and after them. It is said that the black color of subconsciousness represents the darkness of matter with light trapped in it, whereas super consciousness is black light, the ultimate divine light.

As small kids in Iran, we were raised on poetry, especially that of Hafez and Sadee, another master poet of Shiraz, a very beautiful city in southern Iran. Our first exposure to Hafez' poetry, however, was not in school, but rather in local grocery stores. As I remember, a favorite kid's snack those days was grinded chickpeas mixed with sugar and packed in a small plastic bag. Sometimes, if we were lucky, we would find a small rolled up piece of paper in it, about 2 by 3 inches. It was called *fal-e Hafez* and it contained one of his poetry pieces, about 7 to 10 lines written in a very small black font.

It still amazes me how grown-ups entrusted little kids with such a complex poetry that even they could not fully interpret themselves. I suppose, since his poetry has such a magical tone, like a symphonic dance of pearl like words, they thought it would be enchanting to us, and indeed the words were very enchanting. As a little kid, on many occasions, I found those little rolled up poetry pieces in my sugary snack. On one occasion, however, I was really mesmerized by it. Although I could read each word, I could not comprehend the meaning of those lines. But it seemed to me that he was talking about something very important, something everybody should know about.

It talked about angels doing something with humans, not as guardian angels protecting them, but doing something rather experimental and adventurous. I was so taken with that little piece of paper that I decided to keep it until the day I could understand what it was talking about. I thought, for sure I would understand it when in high school. I still remember very vividly its tiny lines written on a small sheet of whitish paper. I do not remember when I lost it, but those lines soon became my favorite poem, even without really understanding them. Some forty-five years later, I have just started speculating about what those angels might have done to us (see p. 301). But why did it take so long for me?

I spent many hours of the long summer days in my teenage years sitting on the windowsill of my room, overlooking the walled front yard, reading Western literature. It was the late '60s and as a nation we had such a reverence for everything Western. The only thing we could think about was looking ultra modern and behaving like Westerners. Like any teenager in any culture, we did not like anything local and traditional. Our food, although very delicious and healthy, looked very different from that shown in the fancy pictures of Western magazines. Our traditional music, although very technical and deep, was too sad and lonesome.

Our poetry books were full of poems so spiritual and unattainable that we could not even comprehend them. Our other classical literary books were written with such technical mastery that every word was ornamented beyond recognition. The typical paintings associated with poetry books looked so medieval and sentimental. Our miniature paintings, although masterfully colorful and done intricately by purposefully not displaying any perspective, seemed unnatural and downright weird. In short, our elaborate Eastern culture, thick and heavy like a black veil, seemed too confiding for our raw and curious teenage minds.

In those teenage years, I read many, many European masterpieces and also some books by great American writers. The last book I read in those years was William Faulkner's *The Sound and the Fury*. The ambiguity of its first part was the ultimate rapture for me. After reading that book, I felt no desire to read any other book as if nothing else could make sense anymore. I remember after finishing each book, I used to write a one-page summary in a special notebook, trying to remember them forever. A few years later, one day while I was a student at the University of Tehran, I went through that notebook again, but I could not remember any of the details of those books. At first, I was so angry with my treacherous mind, for neglecting them so easily. But hours later, I just put the notebook in the trash basket. I thought, if I could not remember them anymore, I must not need them any longer.

Afterwards, I only read science books; little knowing that soon the limited scientific methodologies advocated in these books will confine me in the most fancy mental trap ever.

A Fear Above All Fears

Dying Every Night and Coming Back

Many years ago, my graduate adviser and I were talking about something, now completely forgotten, while slowly moving forward in one of those tiny boats with a mini sail in a lake somewhere in central Iowa. It was a beautiful Sunday afternoon and several people in our lab had decided to go boating in several small boats. All of a sudden, the pole of our boat fell down, causing the boat to turn over. In a short second, just before going down, I fell back into the boat facing the vast sky and suddenly felt surrounded by its majestic cobalt blue color. I was so taken with the sky's presence around me that I just mumbled, "What a beautiful sky," while my poor adviser was struggling to turn back the boat. So for years to come, it became the joke of our lab. He would tell any visitor to our lab that we were drowning and she said, "What a beautiful sky."

Well, it was not the first time that I showed no fear of drowning, and for that matter of dying. I remember when I was maybe 5-6 years old, one day, while picnicking with my family alongside a river outside Tehran, as soon as I saw some grown-ups jumping into the river, I jumped too, and in a matter of minutes I went down right to the bottom of the river. Luckily somebody noticed me and came to my rescue.

Nevertheless, I did not learn my lesson, and some eight years later, it happened again. One day, on a beautiful summer morning, I went to a public swimming pool in Tehran with a high school friend to learn how to swim. Her father was an Olympic swimmer and naturally she was a very good swimmer too. We were the first ones to arrive there. The water in the large swimming pool had just been replaced and it looked like a sea of lapis lazuli, pure and sparkling. Soon I saw my friend up on the diving board, about to dive into the seductive waters. Mesmerized by the water and her swimming act, I stood at the edge of the deep side of the pool and jumped in. soon I was drowning again, and was rescued again, this time by my horrified friend. I still vividly remember going up and down struggling to reach the water surface. However, that experience never left me afraid of water or drowning or death.

Maybe for whatever reason, I did not feel scared to death in the first place to be stigmatized by the event. The horror of death came to me years later, however. Not in the form of any physical or emotional struggle. It came quietly, while sleeping in my own bed many times during a period of about two years.

I am not talking about nightmares by the way, which I know very well. I have had many nightmares; nightmares of living with decomposing dead bodies of unknown people in many different settings. Of the zombie-like body of my deceased grandmother, sitting silently in her grave, with the unbelievable pain of death emanating from her morbid face, as if waiting in vain for her departed soul to come back and bring life to her. I know exactly what a nightmare is, and what it is not. Later, I even learned how to avoid such nightmares, by not eating medium-rare meat at night. But sometimes I sort of miss having a nightmare–the way it jolts my inner world and makes me wonder about the workings of our mind and its capacity to reach unknown mental dimensions.

There is another jolt, however, that I have found less disturbing, albeit very real, a kind of jolt that could be scary to many, but I quite like it. I am talking, of course, about earthquakes. Growing up in Tehran, we were used to moderate, undamaging trembles all the time, especially during summer nights, when sleeping on paved flat rooftops or on wooden beds in enclosed front yards. Maybe I learned to enjoy the motions caused by quakes, either as sudden jolts or rolling shakes; or maybe not, I never could figure it out. While living in Los Angeles and working at UCLA, many years ago, I experienced several moderate earthquakes as well. Indeed, I was always looking forward to experiencing moderate earthquakes. I liked their unexpectedness.

Once, while living and working in Honolulu, Hawaii, I was in a very unique setting when an earthquake struck. My apartment was in a high-rise building, on the thirty-third floor, next to the Ala Wai Canal. It was during the daytime; I was reading while sitting with my back resting on the wall of the living room. Then the building started to rock, moving me back and forth as if being in a giant rocking chair. It was truly awesome and one of those once in a lifetime experiences. Indeed, if I ever get abducted by some outer space aliens, inquiring about humans joyful experiences on earth, being in an earthquake would be right at the top of my list, in addition of being in the presence of smiling little babies with their shining jewel-like eyes staring at me, as well as having beautiful and elaborate dreams at night.

The strongest shake I ever felt was the famous, 7.1 magnitude, Loma Prieta Earthquake in 1989 that shook the San Francisco Bay Area one sunny afternoon on a typical autumn day. It was October 17, right after five o'clock. I was in our lab, standing by my desk writing some comments in my lab notebook.

At the time, I was a postdoc at the Department of Pharmaceutical Chemistry at UCSF. Our building was a huge twenty-two-story modern glass tower. It had a simple, yet uncommon design. The labs were located in the center of each floor, surrounded by hallways on all four sides; and very tall glass panels separated the narrow hallways from a small hilltop forest outside. It was the most beautiful setting I had ever worked in. When walking in the hallways, it was as if one was walking towards a secluded forest.

As I was writing, I suddenly felt as if the building was twisting upward, like a giant tree gathering its roots around itself to uproot, with unbelievable force. While I was in a rapture like altered state, suddenly the tremendous noise of shattering glass broke my concentration. I rushed outside the lab to see how many glass panels had broken. In the middle of the hallway, I saw a graduate student, from a nearby lab, standing in a frozen state as if her shoes were nailed down to the ground, shaking and screaming. I had never seen such a severe reaction to an earthquake. I went close to her trying to calm her down, while thinking at the same time, was she crazy or what? Besides one glass panel falling down, everything seemed okay after a long forceful shake. I suppose we all have our particular fears, and earthquakes must have been hers.

Many years later, I got my share of fear, fear of an awful pain, revisiting me every month for a few days and sometimes for a week or so. It was all because of my right side ovary. Ultrasound tests repeatedly showed no physical problem. So, I was left by myself to ward off a crippling pain with naproxen. What a joke it was, my physician subscribing naproxen for that kind of pain. It would usually strike me in the middle of night while sleeping. Not being able to lay down while in pain, I would start pacing around the dark room, half bended, reaching for the furniture to hold onto, thinking why there should be pain at all, and crying to God, "God help me, let it pass, let it pass." And it would pass, typically 2-3 hours later. Then it would come back the next day and again the next night, for several days in a row. Soon it changed from a severe pain to a terrorizing ordeal of not knowing how many days it would last before vanishing for a few weeks.

Once, when I had my pain, my sister called me from Iran. Having heard my lifeless voice, she wondered what was wrong with me. So I told her that I was in bad pain. Usually, in such cases, with a combination of motherly love and a bit of nosiness, she would give some appropriate medical advice to anyone willing to listen. However, on this occasion, she did not offer any; instead she immediately asked me to pray for her. I thought, how uncharacteristic of her to act so selfishly. I was in such bad pain, and all she could think of was herself. Then she added that prayers work best when offered by someone in real pain, like when women are delivering babies.

I suppose the rationale could be that only through severe pain, our mind would be at an ultimate state of concentrating all its energy to one point. So, I learned to concentrate by praying during my episodes of severe pain to get distracted from that awful pain. I prayed for others and I prayed for myself to become my normal self again. And one day my pain stopped and never came back again, month after month. Apparently, my body learned to cope with the hormonal changes it was going through as I was naturally aging.

Now I have to say, that sometimes, though rarely, I miss that awful pain. Not the pain itself, but the heavenly tranquility that would come right after it, and would stay for about 15 minutes or so. I was usually feeling so light, as if I was devoid of any organ inside me, as if I could fly away. Sometimes I wonder whether or not the ecstasy of the medieval saints we have heard so much about was also due to the aftermath of their ovarian pain. As bad as my pain was, I never felt, even for a second that death was nearby. It never felt like dying, it was just agonizing living.

From some bizarre prior experience, I knew very well of the horror that accompanies death. During a period of two years, I had several episodes of encountering something like death coming to me while sleeping in my own bed, at about 3 am, when the body actually reaches its lowest temperature. I would wake up in the middle of night; actually not waking up, more like being thrown back into existence, but with no memory, no sense of self. The only memory I had was that I was dead. Then I would have only the awareness that every piece of me was somewhere else, that I came back empty, void. And that was the ultimate terror, the unspeakable horror of losing myself, being lost. Not in space, but in myself. Meanwhile, a voice in me would keep saying, "Where *is* me? Where *is* me?" Instead of asking, "Where am I?" As if I was looking for myself. Then little by little, in the dark, the walls of my room would start surrounding me again. Yes, I was real again. I was back in my room. That was the moment I would realize that there was another entity in me, maybe it was my ego, maybe something else. Whatever it was, I was myself again the moment that it had united with me again.

Unlike severe pain, I never felt any tranquility afterwards, and unlike in a nightmare, I never remembered any actual events occurring. I just felt exhausted and wasted. I never ever want to experience those states of mind again. It was like dying and coming back. No wonder we are naturally so afraid of death. It is not easy to let go of oneself. I could not figure out why I had those death-like episodes during those two years. I know for sure that it was not related to any chemical abuse, such as drugs or alcohol. I am one of those people who have to be careful about what to drink for relaxation. Even regular coffee is too strong for me, so I don't drink it at all. Also no tea or coke for me after about four o'clock in the afternoon, otherwise I am kept awake all night.

Shedding Out the Ego

All I know was that those two years coincided with the first two years of a period of four years, when I decided as a scientist to break away from the scientific establishment. This was the time after I left Lawrence Berkeley Lab (LBL, see p. 20) to devote my time to an unorthodox research project that had lived with me, in my mind, for almost two decades. I had saved some money to buy a house, but then I decided to use it to support myself during that risky scientific rebellion. All my life, I was attracted to the scientific establishment like a moth being attracted to light. Then while working at LBL, I was not at peace with it anymore, and there was no more the thread of scientific attraction to keep me there either.

But it was not an easy decision to make. It was like divorcing myself, divorcing my identity, my ego. After leaving LBL in late 1994, I was so busy doing my research project, working all day from seven in the morning to twelve midnight, secluded in my apartment, that I had no time to mourn my separation from a well-defined straight path that I had laid in front of myself since I was a teenager. During those two years, each day and all day, I was in the presence of my true loves–pure and simple numbers–trying to talk to them, to engage them, extracting from them some understanding of the seemingly chaotic universe we live in. And then little by little, they started to talk to me. Trusting me with their untold stories. Soon I became so immersed in these numbers that I even started to dream in numbers at night. Not in the sense of working with them, but perceiving everything as being numbers. Like numbers fighting with each other.

My research was about collecting and analyzing specific data from some 50,000 people, spanning ten generations. I purposely decided not to use any computer program, in order to give myself a better chance to figure out these numbers. I was interested in the characterization of a particular human attribute that I thought might have a genetic component (see p. 92). I think, as I was so busy with my numbers all day, it was only during night in my sleep that I could come to terms with the loss of my original identity, which was entwined with following an easy, straight scientific path.

While being raised in Iran, I grew up as a very proud and almost spoiled kid. At an early age, I started collecting knowledge, though much of it turned out to be perishable. Understandably, I was always praised for it, at home and school alike. I felt I did not need anybody or anything else, except following my thirst to know more and more all day, everyday. I was quite content with my own world.

Nevertheless, I don't believe I came across as an arrogant individual to other people. My desire to understand people and their situations, by asking them endless questions, made me more or less part of any group. For better or worse, I was always at the center of attention at school, constantly being praised by other students and my teachers alike. So much so that when, at age 15, I decided to change my high school, the school principal took it so hard that we all ended up in a very uncomfortable situation by the time I left that school for good.

Nine years earlier, I had started elementary school by going to her school, named after her daughter who was an accomplished painter, and later became our painting teacher. It was a small private school in a nice modern building, with caring and experienced teachers. By the time my two older sisters and I finished elementary school, they had bought a very big house a few blocks away and converted it into a high school for us. Then a few years later, when I was in ninth grade, a classmate who was at our school in second grade, but had left and came back again recently, took me aside and offered me some advice. She asked me, "What are you doing here? This place is too small for you. You belong in Hadaf, not here." Hadaf was a large private high school in central Tehran, far away from our neighborhood in the north of Tehran. It was by all accounts the best high school in Tehran with very high standards for admission.

At the end of my ninth grade, I asked my mother to get my school records from my current school and enroll me in Hadaf. But my principal and her daughter who was, by then, the vice principal, did not like the idea at all. They told my mom that they had added the high school to their original school just to accommodate me and had gathered the best teachers for their school, so there was no reason whatsoever for me to go to another school. My mom went there two more times and still came back with no school records. Next, my father sent his lawyer, but it did not work either. Eventually the lawyer decided to make up a story that my father was relocating his business to another city and we all had to go with him. So at last they saw no choice but to give in and release my school records.

At the end of that summer, just a week before the start of the new school year, my mom sent us, my sister and I, back to our old school for one last time, with a big basket of red roses to say goodbye to our principal and her daughter. We went there with smiling faces, but by the time we left the school that afternoon, my sister was in full tears, while I just felt mission accomplished.

What happened was that when we entered the principal's office, they had already learned that we lied to them in getting the school records. So they were, understandably, very upset about the lying part. I remember exactly one of the sentences my principal told us. She said, "By lying to us, we feel we have not taught you girls anything at all." So they talked and talked to make us feel bad about it, until my sister broke into tears. Then they let us go. My sister and I walked home quietly, after stopping at a corner bakery to get our favorite pastry. I was so excited; I was going to a new high school after all.

Soon I became a star student there as well. But it was a much bigger school, with some 2,000 students, all girls. Every afternoon, at 4 pm, the school alley would turn into a showcase of the latest models of Mercedes Benz, Jaguar, and other top brands, all chauffeured and waiting for us. All we had to do was to study hard and compete with each other. Of course, we were all teenagers as well, with our hormonal levels reaching new highs everyday. I remember my sister, who could not get admitted to Hadaf, had already decided that she would have more fun taking the public bus with her school friends than being driven to school everyday in a chauffeured car.

Not everybody in my new high school, though, was the daughter of a cabinet minister or an army general, albeit some of them were. I also had several classmates coming from other cities and living with their distant relatives in Tehran, just to increase their chances of getting admitted to the University of Tehran.

After finishing high school, I attended the University of Tehran and graduated from its biology department at the top of my class. Afterwards, I went directly to Stanford University for graduate work, to become a geneticist. Although my mother never objected to my decision to come to the US for continuing my studies, and I had been preparing her for my eventual departure since my high school years, when the time came, she looked awfully heartbroken as if her basic right as a mother was violated. Not only heartbroken, she was also worried. I was just 22 years old.

To put things in perspective, I had to remind her that she had left her mother herself at the age of 17 to come to Tehran from Rasht (a major city in the north of Iran near the Caspian Sea, a region as green and as beautiful as Switzerland) to marry my father, who was 22 years old at the time, and to start living with her in-laws, as was common in those days. But to her that was a completely different story. Her parents had to marry her off, as it was right after World War II, and the Russian soldiers who were still stationed in the north of Iran were routinely raping young girls. Of course, that was half of the story. The other half was that she was always dreaming to come and live in Tehran.

She had told us that when she was a school kid, walking home everyday with her classmates, they would pass a Persian rug store full of beautiful rugs. So, one day, she stopped at the store and bravely announced to her friends that when she grew up, she would go and live in Tehran, and every room in her house would have a beautiful rug, just like those rugs. And surely enough, she got her wish years later.

My mom started as a curious, brave and ambitious little girl. But while her best childhood friend, who was also her female cousin, became a successful dentist, she ended up having six kids by age thirty, albeit she got plenty of help raising us too. As I remember, she was always sort of proud of her beautiful face with large and expressive eyes. However, we, her six kids, were the center, the periphery and every inch of her universe. When we were children, all the time, she used to recite to us a children's poem about the merits of books and reading. It was a long poem about a book being the kindest and wisest friend anybody could ever wish for.

My plans for my future started very early in my fourth grade. Our class was on the second floor of a big house converted into a private school. It was in a big corner room with large windows on two sides. There were about 20 of us in that class, each two sharing one big double-sectioned desk. As I was among the smallest kids in the class, I and two other small kids were comfortably sharing one desk in the front row to see the blackboard. However, in the last hour of each afternoon, when we were scheduled to do our homework, usually our teacher would send me to the back of the class, which had one empty chair.

Of course, the last row in the back of the class belonged to big, chubby boys. But they were indeed happy to see me there. I was the top student in the class and no boys ever challenged my position. Since the first grade, when we used to race every afternoon to finish our homework, they knew I could beat them every time by finishing first. One of them was a really sweet boy, who ranked second after me in the class. After finishing our homework, he would always try to entertain me as an incentive so I would come back again. I remember once he had made for me a set of paper-cut dolls, blue in color, and all attached together at the waist. Then he quietly tapped on his desk to make them dance for me. The two corner windows were coming together around his desk, showering us with afternoon sunshine.

One day in that happy, sunny class, our teacher asked us to write an essay about what we want to become when we grow up. I wrote that I wanted to become a discoverer, but my teacher asked me to choose another profession, patiently explaining to me that first I had to choose a specific field to make a discovery in it. All my life, I tried to keep that idealistic ten-year old kid alive in me, and I think after I left LBL, I started to feel that she might be slipping away from me, all because of my big ego. That is why I think I had so much trouble at night in my sleep, going through so many death-like episodes. I think I knew I was losing myself.

I never thought that one day my ego would hurt me like this. All I wanted to do was to know more, and to do things my way. I remember that for the final sixth grade exam, we had to go to another school for a few days, and join other students in our school district to take one set of identical tests. One of those days, at lunchtime, right after finishing the writing composition test, I was called to the principal's office through loud speakers. I did not know anybody there. After going to the office, I learned that they liked my essay very much and wanted to meet and congratulate me, as usual.

As I was getting out of the office, I saw an arrogant looking small girl, even smaller than me, with 4-5 other girls around her, waiting for me outside. In an annoying manner, she asked me why I was called to the office. After I told her the reason, she put her hands on each side of her tiny waist and told me blatantly, "I want you to know that I am going to be the top student in this district." I never heard of such a thing before, to be the top student in a district. I thought, what a stupid idea. Who cares? I never studied for any title.

Not knowing how to react to her, I just gave her a dirty look as if saying, "Get lost, you tiny miny," and walked away. Indeed, I never even participated in any of my several graduation ceremonies. For me only knowledge mattered and everything else was kind of beside the point. So with regard to my ego, it was not something I acquired by myself, rather, with the best of intentions, it was burdened on me by the enormous confidence that my various teachers had instilled in me.

I remember, in sixth grade, when our exams were all oral ones, my teacher, a very caring and seasoned teacher in her forties, would routinely give me the highest grade, even without asking me a single question. She would look at the students and say, "Testing her is just a waste of time; who wants to take the exam next?" She was the same teacher who once asked me to teach a history session in her absence. There I was, a 12-year old kid, and the smallest one in the class, sitting very comfortably in my teacher's chair, and teaching my classmates, "The Rise of the Persian Empire." Soon word was spreading among my other teachers, and by eighth grade, it was not happening on rare occasions anymore. As soon as any teacher had finished her/his lecture to our class, I would be told, "Farideh, go to the blackboard and repeat it for the class." The lecture could be on any subject: proofing algebra equations or geometry theorems, or describing the heart structure. It did not matter, my task was to listen to it once and then repeat it for the class. And I always took it just as a task; I was neither burdened by it, nor euphoric about it. I never felt any pressure, either. Because as soon as they would start teaching the class, naturally I was all eyes glued to their mouths.

With all the attention showered on me by my teachers, sometimes I wondered why I was not hated by my classmates, at least by those who ranked second and third after me. I suppose it was very generous of them. In middle school, in our writing composition class, soon it became like an unwritten rule: nobody would read their essay until I had read mine first. It was just what they always requested from our teacher.

While going to school in Iran, for some 15 years, I was always put in a leading position. But when I came here to the US, all I heard was, "Be a good team player and learn to compromise," and I never liked it. But eventually I had to pay for it, quietly yet severely in my sleep, during my death-like episodes. As bad as those episodes were, and I never ever want to experience them again, I do not regret that once I went through that type of horror, to die a horrible death at night, just to be awakened the next morning. As Rumi, the 13th century Iranian Sufi poet, says in one of his poems, "When was it that I became less by dying?" Referring to his evolutionary observation that first we were plants, then became animals, and later evolved into humans.

Bizarre Connections

A Mind of Its Own

I was driving around South San Francisco, near San Francisco International Airport. It was several years ago, when the airport was undergoing a major expansion, and several very high half-made skyways interlocking each other were tangling in the sky, seemingly going nowhere. It was an awesome site as seen through my car windshield. I was so taken by that maze of grandiose structures that I wished there was a sketch, a drawing, something showing the exact design for the airport expansion.

It was just a passing thought, and by the time I arrived at my work, I completely forgot about it. A few hours later, I was waiting at the desk of a colleague in another department to obtain a work-related document from him. Soon, out of boredom, I started glancing at the scattered papers covering his desk. One colorful loose page grabbed my attention. It looked like a futuristic metropolis with numerous silvery roads connecting with each other very high in the sky. At first I thought that it was a digital setting for a science fiction movie. Then as I was reading down the page, to my absolute disbelief, it said something like, "The new design for the San Francisco International Airport." I thought, wow, I wished I had asked for something more important, as it seemed like somebody up there was listening to me that morning. Or more probably it was just a coincidence.

On another occasion, while working at the same biotech company, something quite unusual happened again. My cubicle was on the second floor, overlooking the San Francisco Bay. There was a big parking lot, separating our building from the bay, with several mature trees scattered around the parking lot. I never minded those trees before. But one Friday afternoon, I was looking outside and noticed one particular tree that was somewhat obstructing my view of the bay. Out of the blue, I thought, I wish that tree was not there. The weekend went by and Monday afternoon, while working at my computer, I casually glanced outside the window, and was kind of surprised that on that day I could see the bay much better.

For a moment, I wondered what could be the reason, and suddenly I realized that the tree, the one I did not want to be there the previous Friday, was gone. Needless to say, I was very confused by the sudden disappearance of the tree. So, I asked around, and was told that because of the storm on Saturday, it had fallen down and early Monday morning was hauled away. I was stunned, not knowing whether it was just a coincidence, or somehow I had predicted its demise in the form of a sudden wish for its disappearance. I stared at the empty spot dumbstruck, already missing the poor tree.

I forgot all about that incident until last year when a similar event happened. There was a big building on Mission Street in Daly City, just south of San Francisco. It was a vacant '50s style brick building, two or three stories high, with a large, impressive, curved staircase in front of it. For some reason, I always thought it must have been a sort of public recreational facility. I used to drive by it quite often, and since it was surrounded on three sides by beautiful, old and tall evergreens, I always paid attention to it. And because of those majestic trees, I always thought that one day the building would be renovated, as it was not badly run down, just out of style. One morning, while driving by it as usual, I suddenly got so cranky seeing it there that I thought, when are we going to get rid of this eyesore? A few hours later, when I was driving back, I saw a big crane tearing down the poor building. I felt really bad, and I thought, oh my God, am I turning into a vampire? I just hoped that they would leave the trees intact.

It could have been yet another coincidence, except that these sorts of little events keep happening to me quite often. I suppose they happen to everybody, but we just take them as coincidence and do not pay any attention to them, until one day, they appear so eerie that we are left with no alternative but to see them as bizarre connections to another state of mind.

The first time I took it as such was some 10 years ago or maybe longer. At the time, I was living in the Inner Sunset area of San Francisco, in an apartment complex. My living room had several big glass windows completely covering the front wall, overlooking a narrow street going uphill, enabling me to see a big portion of the sky all the time. In San Francisco, it is usually either sunny or foggy, and it is unusual to get several overcast days in a row. But on that occasion, it happened that we had a spell of several cloudy days in a row.

One morning, I was looking at the overcast sky through my window and suddenly I thought, so it is like this everyday in Seattle, as I remembered, out of the blue, my dormitory roommate at Stanford University some 20 years ago, who was from Seattle and once had told me that in her city it was usually either rainy or overcast. Minutes later, I went downstairs to pick up my newspaper at the front door. I looked at the front page, and it could have not been more bizarre than what I read in big black font on the top of the page, over a big picture of enlarged raindrops or something like that. It said, "It is like this everyday in Seattle." Referring to our recent unusual weather in San Francisco. It was so weird and scary, not knowing the secret world of your own mind. Afterwards, I noticed that I had a bizarre connection with the written world, especially with my daily newspaper, the *San Francisco Chronicle.*

My love affair with newspapers started a long, long time ago. That might be one reason for my mind being so sensitive to it. Way back in elementary school, my father would not bring home any newspapers. He thought the daily reports of the cruel acts of some desperate people were too much for our young minds to handle. Then, as it is now, Tehran was a vast city and those days in its southern slums anything and everything could happen and did happen. Jealous husbands cutting the throats of their wives, older brothers pouring gasoline and setting their teenage sisters on fire for losing their virginity, or young guys pouring acid on the face of the girls they liked for marrying someone else.

Instead of newspapers, he would routinely buy us colorful children science books, like the ones about stars and planets, or as I remember very well, a book with several pages on the eggs of animals, in all different sizes and colors. He also subscribed us to a newly published science journal in Farsi called *The Scientist.* I remember, when routinely going abroad for business, he would bring us fancy booklets like those from Bayreuth Opera Festival in Germany, or artful calendars from Italy. I really enjoyed looking at those books and journals those days, but nevertheless, I was dying to read a newspaper.

So at last in middle school, my father let me subscribe to one of Tehran's newspapers. I remember, on summer afternoons, close to 4 pm, I would patiently sit at the steps in our enclosed front yard, waiting for the newspaper to be dropped over the front door. Then I would devour every little black line. I would read domestic news, foreign news, bad news, the art section, science section, comic section, obituaries, and every other part except the editorials. Even today, I do not like to read the editorial section of the *San Francisco Chronicle*. I do not like someone else telling me what to think about different issues.

Whether the bizarre connections we occasionally experience in our daily lives are pure coincidences or as yet undefined ways in which our minds are capable of communicating with their outside boundaries is not a question to be resolved anytime soon, considering the slow pace of breakthrough discoveries in science. The slow pace of scientific discoveries is in part due to the resistance of some scientists to take a chance and allow themselves, at least momentarily, to open up to risky and unusual subjects. I remember, when I was doing my graduate work, I once asked a simple question from a young female faculty member at the University of Georgia, in Athens.

In those days, when meeting new people, my staple question used to be, "When is your birth date?" So, when I asked her that question, she looked me directly in the eyes and said, "Shame on you, how could you be in science, and ask me such a question?" I smiled and stared at her thinking, shame on you, how could you be in science and be so arrogant as if you already know everything? As a graduate student in the US, I learned early on that there were two major taboo subjects in genetics, and it was not wise to talk about them. One was the birth date and the other one, human intelligence. Needless to say, I was fascinated by both of them.

With regard to birth date, at a superficial level, it is easy to assume that it is anti-genetics. But if one really understands the genomic world, one would know that it is so adaptive and versatile that it could accommodate a lot of unusual concepts.

Birth date, as I see it, is not about stars and astrology. It is about the observation that people, born at about the same time, share some physical and behavioral characteristics. Of course, we resemble, first and foremost, our parents and close relatives. No question about it. But to understand how the birth date could have some additional influences, we have to remember that in the last several million years, living beings have lived mostly in a predictable changing environment that could have reshaped their genomes through the specific reshuffling of various DNA segments.

The key point, in my argument, is the "predictability of seasonal changes," a notion that seems to have escaped the attention of many geneticists. Due to the fact that seasons are cyclic, genomes should be able to anticipate them beforehand, and respond not randomly but directly. Although this concept, like the theory of natural selection, relies primarily on changing environments as selective pressures to shape the genome, it is quite different from it mainly because in natural selection, the major focus is on sudden environmental changes and not constant seasonal variations.

Nonetheless, some, now mostly forgotten, population studies carried out originally in the 1940s with tiny fruit flies had evidently indicated noticeable seasonal changes in the frequencies of their third chromosomes having two different gene arrangements. Furthermore, according to the natural selection principle, nothing is predictable and solely depends on chance events. Thus, mutation, as the most basic evolutionary force capable of bringing genomic adaptation to a new environment, by definition is considered a random event.

As indirectly mentioned in the previous section, I did extensive research on the genetics of the birth date by studying some 50,000 people in three successive generations. Birth date as a potential genetic attribute, however, like many other traits, does not follow the simple rules of Mendelian genetics. Otherwise geneticists would have figured it out a century ago. After I finished my self-sponsored research, I summarized the results and my tentative suggestions on how it might operate at the genomic level in several scientific papers. But I never could publish them because as I mentioned before, the subject of birth date is the biggest taboo in genetics.

If one plans to be part of the scientific establishment for a long time, one never should mention this word. Ironically many scientists, like non-scientists, when in non-scientific settings, quite often themselves wonder about the role of birth date in defining one's personality. However, since scientific establishment has not yet officially defined birth date at the genetics level, talking about it would be perceived as being anti-science. After all, science, like many other human endeavors, is basically a cult of followers. Just follow the well-trotted path. Genetics, being a relatively new scientific discipline, is still a vast frontier ripe for many new excursions. But in the microcosm of each lab engaging in the characterization of structure or function of one or a few related genes, many molecular biologists and geneticists alike tend to evade the lure of looking for the big picture in this discipline, in favor of more mainstream projects with better chances of receiving scientific funding, and easier gratifications down the road.

Serendipity and Science

Going back to the subject of bizarre connections, although currently it cannot be scientifically defined, its presence, albeit in some other forms, has been well documented in the history of scientific achievements. As many scientists are aware, not all scientific discoveries are the products of following scientific methodologies and sound reasoning. Many of them are simply due to the serendipity factor. Using the exact dictionary description of the word, it states: serendipity "[coined (c. 1754) by Horace Walpole after *The Three Princes of Serendip* (i.e., Sri Lanka), a Pers (Persian) fairy tale in which the princes make such discoveries] **1** a seeming gift for finding something good accidentally…." (From Webster's New World College Dictionary, third edition, 1996).

There is a book published in 2004 about the same subject. It is called *The Travels and Adventures of Serendipity*, written by Robert K. Merton and Elinor Barber. Here, I will quote a few sentences directly from a book review published in *Science* magazine (9 April 2004, volume 304, p. 213). The reason for doing so is to assure the reader that I am not expressing my own views about the role of serendipity in science. The following sentences in parenthesis are direct quotes: "Perhaps the most famous is Alexander Fleming's discovery of penicillin after accidental contamination by a mold dissolved the bacterial culture on which he was working. Others might simply have thrown the contaminated culture away, but Fleming, who had been investing ways to fight infections, went to lay the foundation for the development of antibiotics. The example also demonstrates that as Lois Pasteur said, chance favors the prepared mind. Indeed, only those who possess wisdom can truly profit from luck."

The book also shed light on the cult-like nature of scientific establishments: "Scholars in the science and humanities alike communicate the results of their research in a manner that conceals the tentative nature of the discovery process." And the authors conclude by saying, "Academic discourse hides serendipity's central role in the formation of new knowledge."

Considering the definition of science, which is, "search for natural explanations of observable phenomena," implying that supernatural phenomena has no place in science, it is easy to understand why scientists would rather not talk about the role of serendipity in their discoveries. It seems this definition of science is only applicable to technology minded and goal-oriented scientists (such as those with biomedical bent), and has no operative reality in the search for knowledge. What is the demarcation between natural and supernatural anyways? It seems rather that the former is about the phenomena that have been studied by scientists, and the latter, about the phenomena that have either been neglected or not been studied yet.

The third part of this book deals indirectly with the sudden rise of human intelligence, and at the end, I have proposed the most bizarre theory about the origin of the human mind or at least a part of it. It came to me out of the blue and is indeed a very irrational thought, with no scientific foundation. But considering human myths and our collective memories of the past, it is not unprecedented at all, although it is uniquely original. Had I been told two years ago that soon I would develop such a bizarre idea, I would have laughed my heart off, but as I have reviewed my background in the previous chapters, it seems that all my life I was getting prepared for such a deviant perception of our natural world.

Whether it is right or wrong, only time will tell. The current state of our genetic knowledge is by no means at a level to either prove or disprove it. My decision to publicly express such a bizarre view stems from the fact that being currently out of mainstream science, unlike many other scientists, I have nothing to lose by putting myself in the line of fire. Another reason why I am willing to mock myself like this as a scientist is to stir some controversy in this stalled scientific discipline, hoping to entice others to participate again in this worthy dialogue and eventually reach a more realistic explanation for the sudden emergence of human intelligence.

I owe it to genetics; I owe it to the human mind.

PART THREE

FROM DNA TO MYTH, IN SEARCH OF OUR ORIGIN

The technological advances of the West in the last two centuries have masked the genetic identity of the global Everyman of the 21st century. Indeed, humans, as a species, are genetically more similar to each other than most other animals are similar to their own kind.

In the first half of this part of the book, some common invisible threads, such as our language, fossil bones and DNA, which connect us to a mostly unknown evolutionary past, are pulled out from the fabric of our tainted history and prehistory to find out more about who we are genetically amid our apparent cultural differences. However, since who we think we are may not be quite the same as who we really are, in the second half, some of our ancient myths are brought back to the spotlight to ponder about our current prevailing notion concerning the origin of the human mind.

Thus, an unlikely host of mythical characters, from *divs* to Adam and Eve, to the First Man of the creation myth of Mani, are put into the crucible of our prehistoric time to see whether their amalgam could reconstruct an ancient notion, known spiritually as "divine spark," as the origin of the human mind. In search of a physical counterpart for the incomprehensible divine spark, the enigmatic and ancient Iranian concept of *farr*, vaguely related to the human intellect, is further examined. The intent here, nevertheless, is to remain open-minded with regard to the origin of our vastly mysterious mind.

Who Are We?

Our Changing Languages

The Story of Shaangy

Felix the Cat was my favorite cartoon show some 45 years ago, while growing up in Iran. Every time I watched it on TV, I was dying to know what Felix and the professor were arguing about all the time. However, about three years ago, in 2004, after living in the US for some 29 years, I was so bored with the American TV programming that I started to surf different TV channels, until one day I landed on a Chinese channel called KTSF, broadcasting from San Francisco. Actually, it is an international station with Chinese programming from 7 to midnight every night, except on weekends. Soon, my main connection to American TV became World News Tonight with Peter Jennings, and it became even more sporadic after his untimely death on July 2005.

My interest in the Chinese channel started rather nostalgically. One day, I saw images on TV with some Chinese guys flying around in period costumes. It reminded me of my lazy summer afternoons growing up in Tehran, when our aunt used to keep us quiet by telling us fascinating fairy tales, so grown-ups could take an afternoon nap after eating the customarily big lunch at midday. Most of the time the tales were from the *One Thousand and One Nights* legend. The medieval tale of Shahrzad the Storyteller, a legendary Iranian woman living in the royal court of an Iranian king, who after spending just one night with the king was supposed to die the following morning, due to his crazy antics. So Shahrzad came up with a clever scheme in which she kept the king interested in her ongoing tales night after night for one thousand and one nights, when at last the cruel king fell in love with her and decided not to have her killed.

Another renowned Iranian woman was Momtaz Mahal, who married Shah Jahan, the Muslim king of India in the 16th century. After her untimely death at age 38, during the delivery of their fourteenth child, the king gave orders to build the legendary Taj Mahal over her tomb in Agra, India. Even today, with all current social restrictions imposed on Iranian women, they are still traveling far and doing well these days. For example, there are an unusually large number of Iranian female dentists presently working in the San Francisco Bay Area

Going back to the Chinese TV channel, I soon realized that it offered very good coverage of international news every night, first at 7 to 8 pm in the Cantonese language, and then from 10 to 11 pm in Mandarin. Although I did not understand a word of either language, as it is said, a picture is worth a thousand words. Besides, little by little, I had started losing my interest in the understandable spoken world, typically full of propaganda.

Also in San Francisco, where one in every four or five people is of Chinese descent, it did not seem too unusual to me to switch to Chinese programming. However, even after watching this channel for over a year, I did not develop any curiosity by actually sitting and following a TV series for an hour or so. How could I? I did not understand one word of it. Still, every evening I would keep the TV tuned on this channel. As a die-hard geneticist, it was informative enough for me just to watch people with different facial features and colorful period costumes, speaking strange languages, while doing my evening routine between kitchen, living room and the Internet. That was until December 2004, when I suddenly became a full-fledged Asian by starting to watch and follow one of the drama series of KTSF. For three months, night after night, I watched 70 one-hour episodes of a Chinese-speaking drama. It was basically a drama about women in high power screwing up each other. It reminded me so much of my four years working at a large biotech company in the San Francisco Bay Area (see p. 18).

The main character in this palace drama, with its vast collection of ruthless individuals, was a courageous, hardworking, and talented low class palace girl, who ended up becoming the king's personal doctor about 500 years ago. Until I watched this drama, I had never experienced such torture of not knowing what was going on. Not understanding one word of it, all I could see was beautifully shot and choreographed images with heartbreaking music playing in the background, in which a bunch of people dressed in colorful costumes, with elaborately unique hairdos for women, and weird fanciful hats for men, hidden behind their orderly manners, were ripping each other alive. Caught in their living hell was the young angelic heroine, with a most luminescent face, determined to follow her own path and not get herself entangled in their deadly web of power and greed. In addition to her iron-willed determination, she was occasionally getting some support from her handsome, upper class, idealistic and devoted admirer that made the drama even more engaging.

Originally when I started watching this series, I thought that I was watching a Chinese drama. Players were talking in Mandarin and it had also Chinese subtitles. Albeit initially, I wondered why the palace women were dressed in Korean outfits. All I knew from the *San Francisco Chronicle* TV listing was that it was called *Jewel in the Palace*. It took me several episodes until I realized that their conversations were actually dubbed from another language, but done very professionally. Only several months later after the drama had finished airing, I learned from the Internet that this Korean drama was sweeping across Asia with an unusual appeal across all age groups with its simple message of hope and self-reliance. While watching this drama in Mandarin, I got to know the Chinese names of most characters, and later learned that the original Korean name of the main character was Jang Geum that translated into Chan Jin in Mandarin.

Interestingly, when I was watching some episodes of this drama with my sister, who was visiting me from Iran, we used to call her Shaangy. Because in Farsi, the official language of Iran, we use *sh* and *g* a lot and also like to say *aa* instead of *a*, as well as preferring to end words with *y*. It seemed very fascinating to me how easily one word could evolve into such different sounds according to some regional preferences for pronunciation. Even more interesting was the fact that her original name, Jang Geum, sounded so foreign to us, who became very comfortable with calling her Shaangy.

Now here we are, on this planet Earth, about six billion human beings speaking in some 7,600 different languages and dialects. The difference between a language and a dialect, however, is not always as straightforward as it may sound. It is said humorously that a language is a dialect with army and navy, implying that the decision to officially designate a tongue as a language or a dialect has rather become a delicate political issue in order to underline the unity among the people in some countries. For example, India, with a population of 1.1 billion people, has some 20 official languages, whereas China, a country with a much larger area and about 1.3 billion people, comprising 56 official ethnic nationalities, has only one official language. It should be noted that while Chinese people speak in many tongues, all use the same written language, so in a sense one can argue that indeed all Chinese can communicate with each other through one single written Chinese language. This feat is possible because the Chinese written characters, like Arabic numerals, convey the same meaning while being pronounced differently in different tongues.

In the Steppes of Central Asia

One might ask why should language be used as a tool for finding who we are genetically? The answer is quite simple. Because in most cases we inherit our language from our parents. So tracking our genetic identities through different languages seems to be as worthwhile as searching it through our ancient bones and stone artifacts, or digging deep in our mysterious genomic DNA.

At the same time, languages erode much faster than our DNA goes through changes. It is estimated that by the year 1500 AD, there were already some 15,000 languages spoken around the world. It is also estimated that a new language could arise within 500 to 1,000 years, and it takes about 6,000 to 10,000 years to be completely changed into another language. Despite the fast turnover rate of languages, the science of linguistics has greatly influenced our understanding of the origin of some human ethnic groups.

One prime example is the studies carried out on Indo-European languages. The similarity between Farsi–an Iranian language, Sanskrit–an Indian language, and English has led to the scientific speculation that Iranians, Indians and most Europeans had some recent common ancestors living somewhere in Central Asia. I should mention that currently the name Central Asia is commonly used in reference to only five countries: Kazakhstan, Kyrgyzstan, Tajikistan, Turkmenistan, and Uzbekistan. As I remember, when growing up in Iran, in our history classes and also in our old literature texts, we used to call these countries, which were for a period of about 150 years part of the former Soviet Union, not Central Asia but Turkestan, the land of the Turks.

The similarities between some basic words in Farsi and English, both members of the Indo-European languages, are sometimes quite astonishing. For example, in Farsi we pronounce the word *name* as *naam*, *mother* as *maadar*, *father* as *pedar*, *brother* as *baraadar*, *daughter* as *dokhtar,* and so on. Interestingly, the speculation that Iranians, Indians and most Europeans share some common ancestors going back about 4,000 to 5,000 years fits very well with Iranian and Indian mythologies as reflected in their old religious texts.

In Avesta, the religious book of Iranian Zoroastrians (and other Zoroastrians worldwide), and Vedas, the most sacred scripture of Hinduism, it is stated that the origin of the people calling themselves Aryans (*Aryai* in Farsi) was somewhere in Central Asia, and several thousand years ago they started migrating south to the Iranian plateau, and then gradually split into two main branches, one settled in Iran and the other kept moving until it reached the Indian subcontinent. In the ancient hymns of the Rig Veda, considered the oldest and most important of the Vedas, it is mentioned that Aryans conquered the native people of India and destroyed their forts.

Indeed, this southward migration is believed to be the starting point for the practice of the caste system, still widely common in rural India. The word *catse* is derived from the Portuguese *casta*, meaning "pure" or "chaste." It is not really the best translation, but close enough, for the pattern of social classes traditionally used in Hinduism. Indeed, in Hinduism, it is a very elaborate traditional system made of thousands of subdivisions called *jatis*. One is fixed in a *jati* by birth, and for each *jati* there are sets of rules governing acceptable occupations, foods, marriages, and associations with other *jatis*. In addition, for many orthodox Hindus, this system reflects not only a social order but also serves to express divine justice as either a proper reward or punishment for the deeds of souls in their previous lives.

There are several theories about the origin of the caste system in India. It may have been on the basis of color. The Aryans were supposedly light skinned and it is said that blonde hair was an essential attribute of the people of the highest caste, *Brahmins*. Indeed there is still an expectation in India that higher caste people will have lighter skin, although there are many exceptions, especially in the south of India. The invading Aryan people, according to Rig Veda, had already had a well-established social structure, consisting of three classes. The *Brahmin* was the upper class consisting of priests and teachers, followed by the class of rulers and warriors; and then there was the third class made of merchants, farmers, and craftsmen. The indigenous people, who had darker skin, and most probably spoke Dravidian language, thereafter became laborers serving other classes.

The laborers, now known as *sudras* in Hindu, are not traditionally allowed to learn Sanskrit, or study the Vedas and perform Vedic rituals. And even below the *sudras,* there is another class of numerous *jatis* that are considered by higher *jatis* to be outside the caste system altogether. They are outcastes, and are commonly called untouchables or polluted laborers. Currently there are over 100 million people in India who are designated as outcastes. Some outcastes, called "unseeables," are regarded as being so polluted that they are supposed to keep out of sight and do their work only at night. Many outcastes have converted to Islam, Buddhism, and Christianity to escape the caste system, but have ended up having their own sub-castes.

The Buddha, Siddhartha Gautama, an Indian Hindu prince, founded Buddhism around 2,000 years ago. His renunciation of

Hinduism and establishment of a new religion is believed to have been a revolt against the caste system. More recently, in the last century, Mahatma Gandhi worked very hard to improve the living conditions of the outcastes, or at least to change their status of being routinely classified as non-Hindus.

Although, since childhood, like many other people, I was aware of the concept of the caste system in India, I never knew its precise manifestation in everyday life, until I met a scientist from India some years ago, while working at Lawrence Berkeley Lab. She was married to an American guy and was using his family name as her own surname. One day, out of curiosity, just to hear the sound of an Indian family name, I asked her about her own family name. She laughed rather uneasily and said, "You know what, I don't have any family name of my own." Then, noticing my puzzled look, she explained to me that since her family was from a lower caste in India, they were not entitled to have a family name.

As the main objective of the caste system seems to have always been limiting the cross breeding between different subcastes, its practice has invariably helped to increase the ethnicity within the Indian population.

Going back to our discussion about applying linguistics to find out more about the origin of the human's multi-ethnicity, rather than briefly describing the different languages of various people across the globe, as it is customarily done, here I deliberately decided on another course. The reason is rather simple. Focusing on the people of just a limited number of countries, living in central and western Asia, is meant to discuss more thoroughly some cultural and historical events that I am personally familiar with. So hopefully through this unique approach, it would be easier to realize the extent of the difficulties scientists do have in teasing out what actually happened to us thousands and thousands of years ago. As expected, these difficulties have already created some disagreements among linguists themselves about the origin of many of our current languages. For example, there are some linguists who do not agree with the assumption that Indo-European languages originated in Central Asia. One reason being that currently, among about 50 million people who live in Central Asia, only about 10% of them speak an Indo-European language, that is, Tajiki, which linguistically belongs to the Iranian family of languages.

The rest speak several other languages, all of them members of the Turkic family of languages. The people who speak these languages are mostly descendants from the Turkic nomads who migrated to Central Asia from further east, the areas around the Altay Mountains, where Mongolia, China, Siberia and Kazakhstan all come together. They supposedly moved into Central Asia during several waves of migrations in the past several thousand years, probably traveling through the northern branch of the ancient Silk Road.

Although Farsi is linguistically very different from Turkic languages, in Iran and especially in Tehran we are very accustomed to hearing Azari, one of the members of the Turkic languages. Probably, that is the reason why many people in Tehran know how to swear in Turkish. Some 20 million people in Iran, out of a total population of about 70 millions, speak Azari, the language of Iranian Azerbaijani people. After Farsi, Azari is the most common language, among some 75 languages and dialects spoken in present-day Iran. Most of them are linguistically very close to Farsi. After the Arab invasion of Iran about 14 centuries ago to propagate Islam, Farsi was greatly expanded, and now contains many Arabic words.

Nevertheless, *Shahnameh*, the Iranian epic poetry book written by Ferdowsi, about 1,000 years ago in Dari, the original Farsi form, is completely intelligible to any contemporary Farsi speaker. It is not at all like reading Shakespeare for an English speaker. It is stripped of any Arabic words, which makes it even easier to understand. It is indeed a reminder of how one man's monumental effort and the subsequent support by a local government kept a language alive in its original form after a grand scale foreign invasion that lasted some 300 years.

The Arabic language, which is still widely spoken in some areas in southern Iran, is linguistically quite different from Farsi. Although Farsi uses a simpler version of the beautiful Arabic alphabets, Arabic, like Hebrew and Aramaic, belongs to the family of Semitic languages, which are part of the Afro-Asiatic languages. After the Mongol invasion in the 13th century, some Turkic words were also added to Farsi, including the very common word *khan* meaning "Mr." in Farsi.

When I was going to school in Tehran, almost everybody's grandparents in my class had come from another region of Iran and from a different ethnic group. Therefore, marriage among different ethnic groups, speaking both Farsi and their mother tongues, fast became the norm in Tehran, albeit within the same religious community. Those days, I always assumed that the Iranian Azerbaijanis and the Turkish people of Turkey were genetically very close to each other just because they spoke the same language. So I was very surprised when I recently heard from one of my cousins, who is half Azari, that when he visited Istanbul, in Turkey, a few years ago, he could not even understand Turkish language as spoken by the people of Turkey. That is how fast languages change.

I was even more surprised when I recently learned something quite interesting about the history of Turkey and its people. Maybe one reason for my great astonishment was because, I assumed, I knew at least the basics about the history of Iran's neighboring countries. What I remembered the most about this ancient country was that Turkey and Iran were always at war with each other, whether in ancient times when it was called the Kingdom of Lydia of Asia Minor (currently called the Anatolian peninsula) or as the Ottoman Empire. This vast empire emerged from Turkey and spread across several

Middle Eastern countries and even moved deeply into southeastern Europe and North Africa, some 500 years ago.

So I always assumed that the Turkish people who live in Turkey and speak Turkish were the original Turkic people who later, during the conquests of the Ottoman Empire, also spread to Central Asia. Little did I know that the original inhabitants of the Anatolian peninsula apparently spoke a Proto-Indo-European language, and as late as less than one thousand years ago, Seljuk Turkic armies coming from Central Asia conquered their land. Today, about one third of the people of Turkey are the descendents of the invading Turkic people from Central Asia. However, the rest are from many other ethnic groups who have lived there, some since antiquity and some later on, during Turkey's long history. Some of these ethnic groups, such as the Kurds, speak their own non-Turkic languages (Kurdish is a member of the Iranian languages), and some, whose original languages had long been extinct, now speak Turkic, without being Turks.

So it should be very clear that using linguistics as a major tool to search our human genetic diversity, at times, could get quite problematic and mislead those geneticists not intimately familiar with the history and folklore of the inhabitants of various regions. Azerbaijani people, including Iranian Azerbaijanis, also seem to have a complex ethnic history. Ironically, they may or may not be of Turkic descent. Some limited recent DNA studies carried out among the Azerbaijanis of the Republic of Azerbaijan have shown that they are genetically more similar to the various human populations living in their neighboring countries in the Caucasus region than to the Turkic people of Central Asia.

Even more fascinating is another Iranian ethnic group called Taleshi, living in an area called Talesh, tucked away between Iran's Azerbaijan province and the Caspian Sea in the north of Iran. Although these people have their own language, which is widely spoken in both Talesh and the adjacent Republic of Azerbaijan, in some areas of Talesh, as in its major city Astara, some of these non-Turkic people speak Azari as their mother tongue. However, these people strongly object to being called Turks.

Some Iranians, especially the people of Tehran, who now live in a vast metropolitan area of over 10 million people, have great difficulty in understanding their linguistic predicament. For example, a few months ago my niece, being unfamiliar with the intricacies between ethnicity and language, while talking to me on the phone, sounded quite frustrated that a Taleshi friend of hers who speaks Farsi with a thick Azari accent still adamantly denies being Turk.

Now let me tell you a little about Astara, the beautiful border city of the Taleshi people, neighboring the Republic of Azerbaijan in the south of Russia. It is situated in a forest like setting, and I still remember its dreamy deep jungles, where, as an undergraduate student at the University of Tehran, many years ago, my classmates and I visited on a field trip to learn firsthand about wild mushrooms. That was the first time I saw a jungle and also the first time I saw a border between two countries. But it was not at all like what I thought that a border would look like, just a small narrow bridge over a sleepy quiet river, where along its banks on both sides a few guys were idly fishing.

Just a few feet away, on the other side of the bridge, behind a large iron door, typical of those seen often in many parks, was the mighty Soviet Union (that was before the Republic of Azerbaijan became independent). That scene, with all its simplicity, seemed so surrealistic to me. I think that was when I realized that no matter what our nationality or ethnicity, we are all just a bunch of people caught in our own history.

Going back to the Turkic languages, linguistically speaking, they are part of the Altaic languages, which are spoken mostly in and around Central Asia and Far East. Altaic languages also include the Mongolian language and some other languages spoken in Siberia and Manchuria. Some linguists also prefer to include Korean and Japanese among Altaic languages.

At first this suggestion seems very unrealistic because we are all accustomed to bundling Chinese, Japanese and Korean languages together, as the people who speak these languages have many similar facial characteristics. However, after switching my TV channel, a few years ago, to KTSF, which has mainly Oriental programming, soon I realized how much Oriental languages sound different from each other. For example, Mandarin, a Chinese language, and the Korean language sound distinctly different. Indeed, to me, Korean sounds much more like Turkish than Mandarin.

Furthermore Korean women, with their round and full faces and high cheek bones, do look kind of similar to Turkish women of Central Asia, who also have slightly slanted eyes, and in some cases puffy eyelids. After all, since ancient times through the northern branch of the Silk Road, traveling the long corridor south of Siberia between the Caspian Sea and the Pacific Ocean, these two seemingly distant groups of people were in regular contact with each other.

Now let's go back to the Indo-European languages and their single representation in Central Asia, the Tajiki language, the official language of Tajikistan. Linguistically speaking, Tajiki is grouped among Iranian languages, and speakers of Tajiki and Farsi can understand each other quite well. Indeed, I have heard that Iranian singers used to be very popular in Tajikistan. A small country of about 7 million people in Central Asia, Tajikistan is situated north of Afghanistan, between two great rivers called Amu Darya (historically known as Jayhoon) and Syr Darya (historically known as Sayhoon). This region was known as Transoxina in ancient times. Notice that the word

Jayhoon (Oxus in Greek) sounds very similar to Gihon, one of the four rivers of the Garden of Eden according to the Old Testament.

Although Tajikistan and Iran do not share any border with each other, they share a long history. Historically, people of Tajik descent have always been grouped among Iranian people. In our history books, we called the present-day Tajikistan, Sogdiana. During the Achaemenid Empire of Iran some 2,500 years ago, it was one of the states of the Persian Empire which paid tribute to the central government located in Persepolis in the south central region of present-day Iran. However, at the end of the Sassanid era, some fourteen centuries ago, when the gates of the Sassanid Empire fell under the invading Arabs who brought Islam to Iran and beyond, it was eventually Sogdiana that revived the ancient Iranian culture and language during Arab dominance. The Samanid Empire emerging from Sogdiana ruled over Iran for a century and played a fundamental role in preserving the Iranian culture, as we know it today.

Due to the entangled history Iranians share with the Tajik people, now the two people not only share the same language, but also many customs, including an elaborate 5,000-year-old New Year ceremony called *Noruz*, which celebrates the coming of spring. I should mention that *Noruz* is celebrated in other regions of Central Asia and the Caucasus as well. The Iranian calendar still starts with *Noruz*, a Zoroastrian tradition, which makes other Muslim countries quite wary of the Islamic Republic of Iran. But let's face it, in the language of nature, spring is the only time when each year starts to renew itself.

Although Tajikistan is now far away from Iran, Iranian classical poems written centuries ago, but still widely recited, are filled with longing for Samarkand and Bokhara, two legendary ancient cities located along the Silk Road, which are still major cultural centers of the Tajik people. Ironically, these two cities, inhabited mainly by Tajik people, are now part of the neighboring Uzbekistan, and not Tajikistan. Samarkand, with its outstanding Islamic architecture, a city full of blue mosques, goes back 4,000 years, and is one the oldest continuously inhabited cities in the world.

It should be clear from the few cases mentioned in this section, from Turks to Tajiks, that through the tilted lens of human regional history, what scientists would see a lot of times is not the real ethnic identities of different human populations, but in many cases their imposed mother tongues and transplanted traditions. It takes a lot of regional understanding to find the invisible threads that connect specific groups of people together.

Now let's further examine what could be the reason that Tajikistan is the only country in Central Asia where an Iranian language, that is, an Indo-European language, is still spoken, while all surrounding countries speak Turkic languages, not belonging to this family of languages. The key point seems to lie in the geography of Tajikistan. This small country is covered with very high mountain ranges. Indeed, the tallest mountain in the former Soviet Union was located in Tajikistan. But this mountainous area is partially surrounded by lowlands, some of them among the harshest deserts on earth, which are occupied by its Turkic speaking neighbors. Apparently, Tajik people, living mostly on highlands for thousands of years, had the unique opportunity of keeping the language and traditions of the original Indo-European people intact, away from various cultural elements brought on by the periodic migration of nomadic Turkic people coming from further east.

Surprising Genetic Findings

Whether the linguistic approach outlined above could justify the designation of Central Asia, as the original place of the ancestors of Indo-European people, including Iranians, will need further clarifications by relevant genetics studies. Limited genetic studies indeed show remarkable genetic similarity between the people of the Middle East and Europeans. Whether their last common ancestors came from Central Asia or the Anatolian peninsula (Turkey) or elsewhere remain to be resolved.

Cavalli-Sforza and coworkers published one of these studies in 1988. In this study, 42 populations from around the world were analyzed for several of their polymorphic proteins. Polymorphism means the presence of two or more alleles (variants, forms) for a given locus (gene), or gene product like protein, in a population. The genetic data were presented as the gene frequencies of these polymorphic loci (plural for locus). Genetic similarity is indicated by similarity in the frequencies of alleles under study. A total of 120 alleles were analyzed among these people.

Using these genetic data, a human evolutionary history was reconstructed which was then presented in a phylogenetic tree (chart). The first split in this tree separates the majority of Africans from non-Africans, and the second split branches into two major categories, one corresponding to Caucasoids (Caucasians), East Asians, Arctic populations, and American natives; and the other corresponding to southeast Asians, Pacific islanders, New Guineans and native Australians. It should be mentioned that this chart is not necessarily an ideal representation of all the people of the world simply because there are many other human populations that have not been included in constructing this chart. But understandably, whatever data were available at the time, around 1988; they were used to construct this genetic tree.

Also, in this chart, some populations are grouped as one; for example, some Europeans are lumped together as one population. While one specific European population, the Lapps (also known as the Sami people of the Nordic region of Europe) represents one category by itself. What was surprising to me was that in this genetic tree, in the Caucasian branch, the closest population to the European population was the Iranian population and next to it was the population from southwest Asia. Although as an individual, this information surprised me, as a geneticist, I was quite expecting it. Because both Europeans and the people of the Middle East, as well as the people of India always have been grouped as Caucasian, mainly due to their somewhat similar facial characteristics.

The closeness of Iranians and Europeans as implied by this genetic tree could be an indication that they had common ancestors not long ago, maybe going back several thousand years, as predicted by the similarities between Indo-European languages. As I mentioned above, as an individual, I was very surprised to see, in a large chart, Iranians and Europeans sitting next to each other so closely. And I wondered why–why was I so surprised after all? Then I remembered that although I have lived in the US for 31 years, mentally I still feel closer to my Asian friends, many of them Oriental, than to many of my American and European friends.

So it seems to me that despite all the talks and interest in genetics these days, what we really identify ourselves with has more to do with our cultural similarities than our genetic attributes. That is probably the reason I could sit and watch Korean, Japanese, and Chinese historical dramas on TV without understanding a word of what they are saying. It is not necessarily their traditions or customs, but their mannerisms that look very familiar to me, and with which I can, as an Easterner, strongly identify with. As a geneticist, I find it rather ironic that a major component of our human ethnicity is rather cultural, and therefore, more transitory in nature. But as I have found out through my training in evolutionary genetics, even the part of our ethnicity that is deeply rooted in genetics is indeed nothing but a transitional state. There is simply nothing about our physical characteristics that could be considered as pure or permanent.

I can still vividly remember the day when, in sixth grade, I read in our history book that the origin of Iranian people is from a group of people called *Aryai* who migrated from southern Russia's steppes (now commonly called Central Asia) some 5,000 years ago (notice the similarity between *Ira* in Iran and *Ary* in *Aryai* which is the same as Aryan in English). I was utterly stunned to read such strange stories about my Iranian ancestry and indeed, I took it very badly as if I had been told that my parents were not my real parents and I was an adopted child. I thought, it's so strange that we came from somewhere else. Before that day in my history class, I had always thought that being Iranian meant that my ancestors had always lived in the geographic location called Iran. I suppose as kids we don't understand the transitory nature of everything around us. We just assume that everything has always been the same way.

Since then, I have always been curious about "those people" who migrated south to Iran so long ago. However, what our history books did not tell us, or maybe I did not pay enough attention, was that while those people were moving into present-day Iran, they encountered some indigenous people and got mixed with them eventually. As a kid, I always thought that "those people" had moved into a land with no human inhabitants. Indeed, the movements of new people into an already inhabited land and the consequential mixing of the two groups could be at the heart of our present-day multi-ethnicity. And considering that not only one but many, many waves of human migration into the same region, and likewise all over the world, could have occurred during hundreds of thousands of years of our prehistory, it would not be so surprising to see present-day humans in so many different shapes and sizes.

Evolutionary Forces and the Myth of Darwinian Selection

A Brief History of Evolutionary Studies

Newspapers with large circulation in typical metropolitan areas regularly write about evolutionary topics. These articles are usually written by competent writers who know how to spark interest and admiration in their readers for 15 minutes or so while they are momentarily taken away from their routine metro train commutes to visit the mesmerizing world of science and scientists. One day the story is about how natural selection, through the principle of survival of the fittest, has shaped human evolution, and another day it is about how African Eve some 50,000 years ago ventured out of Africa and gradually her descendents evolved into different ethnic groups and populated the whole world, as we know it now.

To the general public, these two topics are just versions of the same subject that asserts we humans, like any other organism on this planet, are products of organic evolution. However, as an evolutionary geneticist, many times I have wondered what would happen if one day these two particular topics were written side by side in a newspaper. Could people notice that the evolutionary rationales behind these two topics were in principle contradicting each other? What would their reactions be if they find out that according to our current evolutionary principles only one of these two scenarios could be plausible and not both of them?

To be honest, I have included this section with much hesitation because it is rather obvious that the general public is not interested in the details of evolutionary processes. But without knowing these details, how could we trust what is written for us? Do we want just to be entertained by science or do we want to be able to evaluate it as well?

Since Darwin published his ideas about human evolution in the mid-19th century, the general public has shown great interest in the subject of evolution. Periodically, this interest becomes more intense, indirectly due to a sudden jump in our technological advances. People trying to buffer themselves from these fast-moving changes, imposed on them by a small fraction of influential people at society's top level, naturally seek solace in their religious roots.

If the present is any indication of the past, I think it could be the case not only now, but also back one or two million years ago, during the stone age, when a few innovative/ technical people in each generation or so gradually or periodically advanced human's technical capacity as a whole. During the last two decades or so a new surge in religious interests has brought the subjects of evolution and Darwin back into the spotlight again. Given the recent intense public interest in these subjects, I decided to take the risky task of writing rather in detail about evolutionary processes in this section.

My main purpose in doing so was to give a new perspective to this old subject. As I am not currently affiliated with any scientific establishment, I am not going to give any standard or official view of what evolution has to be. There is more to evolution than natural selection, and ironically there is more to natural selection than Darwin. Indeed, natural selection is a common name for a complex of processes, some of which act in opposite directions in driving evolution.

Since the general public has always equated evolution with Darwin and natural selection, I will first give a bird's eye view of evolution from the perspective of Darwin's natural selection, while bringing into the spotlight other evolutionary forces or pressures. I should emphasize that this overview is not about human evolution at all; rather, it is about evolutionary processes in general, affecting all organisms. Darwin's natural selection is a simple principle that can be summarized in one paragraph. But if some people are really interested in understanding its full scope, then they should try to understand it the hard way, that is, how it relates to other evolutionary forces, including other types of natural selection.

Let's now examine the known principles that are thought to be responsible for organic evolutionary changes, and to do so properly, let's start by reviewing the historical developments of some concepts that eventually shaped our thoughts about evolution during the mid-19th to mid-20th centuries. I should mention that, for the sake of brevity, this is a very brief summary of the events leading to the establishment of the population genetics field. Indeed, it is so short that many essential milestones had to be left out.

As a person trained in evolutionary genetics, I feel an obligation to describe how, historically, many scientists working in this field had dealt with the principle of natural selection. It is noteworthy to point out that natural selection is no longer the central focus of evolutionary thoughts, especially with regard to humans' present-day diverse ethnicity. It is just one evolutionary force among a known group of four. On the other hand, since the general public and popular press has developed a fixation with the principle of natural selection and has come to equate it with human evolution, I am devoting more attention to it in this section, hoping to clarify some misunderstandings about the nature of life, which is as much about consistency as it is about evolving through time. Of course, there are still a few evolutionary theorists out there writing books for the general public, and feverishly advocating for Darwin and natural selection as the driving force of evolution. Despite these efforts, the great enthusiasm of a few scientists cannot redefine the historical course of a scientific field.

Science, like many other human establishments, has always played a big role in cultivating hero-worshipping among the masses, albeit just for a handful of its brightest stars. Although it is very common in both scientific journals and the popular press to talk about Mendelian genetics and Darwinian evolution, Darwin was not the first thinker who elaborated on organic evolution. Prior to him, in the first half of the 19th century, there were other proponents of the theory of evolution. In the second part of the that century, when he first published his ideas, he had been greatly influenced by other thinkers of this period, such as geologist Charles Lyell, the father of modern geology, and Thomas Malthus, an early political economist who had started warning people about the consequences of uncontrolled human population growth in a world with limited natural resources.

The perfect timing of Darwin's ideas along with those of his other great contemporaries could well account for the early public and scientific acceptance of his theories during his own lifetime. Unlike Mendel, the founder of modern genetics, who was way ahead of his time when he proposed the concept of the gene as a discrete unit of inheritance. Thus, in his case, the scientific world had to wait until the beginning of the 20th century, years after his death, to comprehend his visionary ideas and come to terms with his laws of inheritance.

Darwin's most widely known contribution to evolutionary thoughts is his idea of evolution by descent, that is, humans and great apes, going back in time, but not too long ago (in the evolutionary time scale), had a common ancestor, implying that among other animals, great apes are our closest living cousins. Since chimpanzees and gorillas live only in Africa, he also suggested that humans must have evolved in Africa. The orangutan, another great ape, on the other hand, lives only in Southeast Asia.

Interestingly, the common, somewhat romantically nostalgic, pictorial depiction of human evolution, showing our gradual change from a knuckle walking chimpanzee to a handsome upright human, has unwittingly created a great misunderstanding among laypeople about human evolution; the fallacy that the human has descended directly from great apes. Most likely, our common ancestors did not look either like humans or great apes, but somewhat similar to both.

Darwin's other widely known contribution was the development of the principle of natural selection as a mechanism to bring about evolutionary changes, that is, evolution by natural selection. So what is natural selection? Historically, there are two definitions of this concept. At Darwin's time, it was equated with the differential reproduction of individuals, who were carriers of different hereditary attributes. That is, any person who produced more offspring would be naturally selected for, and his/her traits would have the greater chance to propagate in future generations. But to do so, these traits should also make her/his offspring (and their offspring) better fit and able to produce more offspring, until these traits become fixed in the population, and consequently a new species would gradually evolve.

It is indeed a simple and obvious concept, at least for the fittest individual in the first generation, and has been around since antiquity. Empedocles talked about it in ancient Greece. Lucretius talked about it in ancient Rome; even Darwin's own grandfather, Erasmus Darwin, talked about it long before he did. Some scientists even preferred to call it the Darwin-Wallace theory of natural selection because these two naturalist acquaintances and Englishmen independently proposed it and published it around the mid-19th century. It was also well known that, over 20 years earlier, Edward Blyth, another Englishman, a zoologist and chemist, widely wrote and published his ideas about the same concept, without calling it natural selection.

So what was going on around the mid-19th century that generated so much interest in this concept of natural selection? The answer was a growing public fascination with the process of artificial selection. In those days, making new breeds of domestic animals and plants through selective breeding was becoming a common practice. So Darwin, understandably not knowing about other evolutionary forces, advocated his new concept of natural selection as the main mechanism by which humans could have evolved from animals, albeit during a much, much longer period of time than was achieved through selective animal breeding.

Although at the time, there was no clear understanding of the concept of gene as a discrete unit of inheritance, essential to the practice of artificial selection was the general acknowledgement that varieties existed in wild population of animals. Otherwise, if they were all the same, no amount of selection would be effective in creating new breeds of plants and animals. However, it became soon evident that Darwin's theory could not answer two fundamental questions concerning the existence of varieties in wild populations. First, why was there so much variation within a population in the first place, that is, what did cause the varieties to occur? And second, what was the mechanism by which variation was inherited?

Darwin did not have any answer for the first question, and his answer to the second question made things even worse. At his time, the prevalent view about inheritance was that parental traits in each generation would get blended, and therefore, the offspring's traits would represent an average of parental traits. What it meant was that natural selection could not work within the framework of the blending theory. Since a few "fitter" variants in the population would blend with less fit variants in each generation, after a few generations the "fitter" trait would be diluted out completely. And that was how natural selection lost its appeal among biologists by the end of the 19th century.

Meanwhile, during this time, at the beginning of the 20th century, Mendel's work was rediscovered and his "factors," now called genes, were universally accepted as the discrete and permanent units responsible for the transfer of various traits from parents to their offspring. And that meant the end for the blending theory of parental traits. According to the Mendelian laws of inheritance, each gene would sustain its basic characteristics, without blending, during transfer from one generation to the next one. But generally speaking, each gene has to follow an enigmatic natural genetic rule that dictates whether a gene is going to be dominant, recessive or codominant in the next generation. This system of gene expression is akin, at a superficial level, to a relaxed genetic caste system.

One of the three scientists responsible for rediscovering Mendel's old experiments and establishing the new discipline of Mendelian genetics later advanced his mutation theory, based on spontaneous genetic change. This theory soon took center stage and replaced natural selection as the driving force of evolution. Natural selection was pushed back as mainly a purifying agent responsible for the elimination of harmful variations. Although mutation theory enjoyed widespread acceptance for three decades especially among many pioneering and prominent Mendelian geneticists, it soon became evident that mutation alone could not account for much constructive evolutionary changes. It was shown that in laboratory animals, most of the time, mutations caused only deleterious effects.

Today, mutation, either as a random change in DNA structure or a rearrangement in the structure of chromosomes (where genes are physically located in the cell), remains as the only evolutionary force capable of creating variability, whether destructive or constructive, within any genome. But there is more to evolution than mutation because most times a mutation is neither useful nor harmful by itself. Only a specific environment can determine its adaptive value. Thus, if a genotype (the genetic makeup of an individual) carrying a new mutation survives its particular environment, it is said to be adapted to that environment.

After the gradual demise of mutation theory as the main evolutionary force, selection was consequently revived by a handful of geneticists who saw in Mendelian genetics and genes—as discrete units of inheritance—a perfect genetic base for natural selection. No more blending theory to worry about. Nevertheless, there was still a problem with Darwin's original idea. In Darwin's view, natural selection acted only on individuals. Now with the gene at center stage, natural selection only could be perceived as operating at the level of a single gene or combination of genes or a trait itself, but not at the individual level.

That meant that the original concept of natural selection as described above, "the differential reproduction of individuals…" was not operative anymore. So later the notion of individual was discarded and it became: "Any new physical or behavioral trait that enhances the survival of individuals and increases their chances of reproducing more offspring than others, will increase the chance that that trait will have higher frequency in the future generations."

Conceptually, it meant a drastic departure from how Darwin approached evolutionary changes. Rather than a change in individual fitness, an evolutionary change became a change in the gene frequency of its core population. At the same time, as a result of natural selection, genetic adaptation becomes the adjustment of populations, and not individuals, to their environments. Switching from individual to population also means that even if the "fittest" organism survives by leaving relatively more offspring, still there won't be any guarantee that the trait that made her/him the "fittest" will be transferred to the next generation. Each offspring receives only half of its genes from a particular "fit" parent. So in the transfer of a gene responsible for an advantageous character, the chance would be 50-50 that an offspring will receive it.

There are, additionally, other genetic constraints that are beyond the scope of this discussion. There is one situation in which all males can transfer their "fit" genes to all of their male offspring; that is, when the advantageous gene is located on the Y-chromosome, a chromosome that, generally speaking, is included only in the genome of a male organism. However, this chromosome, being the smallest chromosome in the human genome, carries a very limited number of genes. Thus, the chance that a functional gene gets randomly muted on this chromosome is much lower than a mutation occurring in much larger chromosomes.

Thus, when a person dies, her/his wholeness also dies with her/him, and only some of her/his genes get a chance to live in future generations, unless he is Genghis Khan who supposedly left 1,000 offspring behind. Otherwise, given that all genetic requirements are met in every consequent generation, it will require a long, long time for a specific gene to increase from one sample in a given population to complete fixation, that is, to be expressed in all individuals of that population.

Today, some 70 years after this conceptual change from individual to population, many non-scientists and some scientists, who are not population geneticists, would still rather talk about "survival of the fittest" which has no evolutionary meaning anymore and instead only represents a far limited social concept. The transformation of Darwin's natural selection by Mendelian genetics, mostly through mathematical reinterpretations by three scientists, in the late 1920s, became known as synthetic theory or Neo-Darwinian theory. Soon after, two other evolutionary forces, migration and random genetic drift, became incorporated in this theory which became the foundation of modern population genetics.

Around this time, a new development occurred that fundamentally changed the way evolutionary processes were studied in the lab. Up to that time, the common practice was to generate mutant organisms, by X-ray radiation or chemical treatments, and then to study the genetic changes that resulted from these treatments. One reason for this routine experimental practice was that natural mutants were generally viewed as rare occurrences and nature's freaks, and the wild type of each organism was considered to be the norm, the archetype.

But those who were studying natural populations in the wild knew that the wild type was only a laboratory label, and in nature there were many organisms with a wide range of variations in their physical and behavioral characteristics. One of those naturalists turned geneticist was a Russian immigrant called Theodosius Dobzhansky. By collecting variant forms of tiny fruit flies from the wild and bringing them to the lab, he ushered a new way of assessing the actual action of natural selection. I should mention that although evolutionary processes are considered to be very slow and not noticeable during one's life time, under certain environmental conditions, when selective pressure gets very high, it is possible to follow it experimentally and learn from it.

Testing Evolution in the Lab

I remember many years ago, right after I came to the US and attended Stanford University; I used to wonder why scientists write so many textbooks. They all seemed to talk about the same subjects, using the same writing style. Then I took a course in population genetics given by Professor Cavalli-Sforza (originally from Italy) that was supplemented by one of his own textbooks. Using a new approach in addressing different issues in population genetics, as well as his candid style of writing, made for a quite exceptional textbook and also made me a big fan of his.

So one day I went to his office, trying to find out whether I could join his group as a PhD student. But it turned out that as I was a foreign student, my GRE score was not high enough to join his lab, so he suggested to me to join the population genetics group at UC Davis. He knew that Dobzhansky, the Russian naturalist turned geneticist, as mentioned above, had established a big lab there, attracting many graduate students. In those days, he and Dobzhansky were both well-known geneticists. I should mention that Dobzhansky had actually died in 1975 and another scientist was running his lab at the time.

Cavalli-Sforza and his group were working with humans, collecting blood samples from people all over the world to compare the gene frequencies of many corresponding enzymes and proteins among different human ethnic groups. He was interested in figuring out the early migration patterns of prehistoric humans. Dobzhansky and his group, on the other hand, were working with fruit flies, following their gene frequencies under different selective pressures, such as high alcohol concentration, and trying to figure out the details of evolutionary processes.

I did not know then, but later, as a PhD student, I learned that in those days, as a population geneticist, you were either a "selectionist" or a "neutralist" (see below). So in this sense, Cavalli-Sforza was a neutralist, whereas Dobzhansky was a selectionist. Although I did not go to UC Davis to join Dobzhansky's group as was suggested to me by Cavalli-Sforza, I somehow ended up in the lab of one of the people in Dobzhansky's group, who later became my graduate advisor. He was an American of Irish descent and used to tell us many stories about how it was like to study in the lab of a big scientist, who was later remembered mostly through one of his great statements, "Nothing in biology makes sense except in the light of evolution." The one story I still remember was that no matter how late graduate students would work at night if they were not in the lab at 8 in the morning, Dobzhansky would not even acknowledge them.

So in the lab of my graduate adviser, at Iowa State University, for about four years, I was engaged in doing experiments that basically followed Dobzhansky's vision of how organisms have evolved through natural selection. Our experimental organisms were tiny fruit flies, scientifically known as *Drosophila*, and they seemed perfect for this task. They have a short life span, about two months, and could be easily followed for 30 to 40 generations in the lab. As they are highly alcohol tolerant, they are well adapted to live in the wild in rotting fruits, which have high concentrations of different alcohols. They consume alcohol and turn it into an energy source. The major enzyme in the chain of reactions that detoxifies the alcohol is called alcohol dehydrogenase. Since this enzyme and the gene that makes it are arguably the most studied enzyme and gene in any organism, we used to call this enzyme the queen of enzymes. The main reason for this enzyme-gene system being so widely studied was because fruit flies, by living naturally on rotten fruits, are constantly subjected to a very high selective pressure, making this system very suitable for assessing the action of natural selection in an unusually fast forward pace.

Interestingly enough, the gene responsible for this enzyme has not become fixed in the wild yet. It comes in two allelic forms—one produces the enzyme with high activity and the other makes the enzyme with lower activity. The allelic form that makes the high-activity enzyme causes the faster breakdown of the alcohol, and therefore, is more adapted to this environment. So our experimental rationale was that if natural selection were the main evolutionary force operating at the alcohol tolerance trait, one would expect that flies with the high-activity enzyme would be fitter, and therefore, leaving more offspring. That is, the gene for the high-activity enzyme should already have replaced the other allelic form of the enzyme and became fixed in the wild.

However, there was the possibility that the fruit fly adaptation to high alcohol concentration was a more recent event and more time was needed for the complete fixation of the more adapted allelic form. That was why we and other labs were trying to facilitate the necessary conditions to see whether we could accelerate the process of allelic fixation. Describing the many details of these experiments and their results is beyond the scope of this book. I brought up this experiment to highlight how it is possible to study evolutionary processes in the lab. I should just mention that the results were not as simple as we originally thought, and because of that we learned much more than we expected from these experiments.

To briefly summarize, one of the results that I obtained indicated that the gene for the low-activity enzyme persisted in the wild because in homozygotes, that is, in flies with two copies of this allelic form (one from each parent), it made them better adapted at very low and very high alcohol concentrations. But in homozygotes with two copies of the gene for the high-activity enzyme, it made the flies better adapted at medium concentrations of alcohol. Yet those types of results were not interesting any longer for the other scientists in this field.

There were three major reasons for this latest transformation in the course of evolutionary studies. First, it was just another case of balanced polymorphism (see p.135). In those days, it was becoming very common to get this type of result when doing selection experiments. Second, in the early 1980s, with the advent of DNA cloning and recombinant techniques, working with DNA seemed more exciting than working with enzymes and proteins as we used to do, and many population and evolutionary geneticists started to design new studies applying the new molecular biology techniques, and soon the Maniatis *Molecular Cloning* voluminous lab manual became the new gospel in every lab.

Third and most importantly, the population geneticists became weary of the constant debate between themselves, that is, selectionists versus neutralists ideologies. It seemed like one of those contests with no clear winner, ever. The selectionists, as I mentioned above, were those advocating for natural selection as the driving force of evolution.

So let's see, who were the neutralists? Beginning in the 1950s, some mathematicians turned population geneticists started to argue mathematically about what would happen if a new mutation would not make any change or make a very small change in the function of the gene it affects. That is, it becomes neither useful nor harmful to its host. In other words, it would act neutrally. Later, it turned out that indeed, some genes with new mutations did not give any selective advantage or disadvantage to their hosts, regardless of the specific environmental conditions in which they had arisen.

That was when many population geneticists started to have new doubts about natural selection and its role as the driving force of evolution. It became plausible, therefore, that some new mutations with no adaptive value would become fixed in the population just by chance. Thus, generally speaking, in the last two decades, natural selection again has lost its luster among many geneticists. More recently, studies with mitochondrial DNA and Y-chromosome DNA have also indicated that these stretches of DNA are not under any selective pressure. That is exactly why they are currently being used as tools for genetic identifications (see Our Mischievous DNA on p. 165).

Different Types of Natural Selection

Consistency vs. Change

Despite the failure of population genetics studies to legitimize the notion that natural selection is the driving force of evolution, population geneticists learned considerably from those experiments. The most important thing we learned was that natural selection is a common name for a complex of various processes. There is balancing selection, different types of stabilizing selection, and directional selection.

Directional selection is also called Darwinian selection because it is historically the only type of natural selection that was described by Darwin. Directional selection generally operates in response to a sudden environmental change and selects for the most extreme characteristics in a population. So it is likely to result in the gradual replacement of one allelic form of genes with another. It is also the selection regime exercised by plant and animal breeders seeking to enhance certain traits.

On the other hand, balancing selection favors genetic polymorphism, that is, the maintenance of two or more allelic forms of the same gene either in homozygotes (those who have inherited two exact copies of the same allelic form of the gene, one from each parent) or heterozygotes (those who have inherited two different allelic forms of the same gene, one from each parent). As described above (see p.130), alcohol adaptation represents an example in which both types of homozygotes are favored—in this case, the flies with high-activity enzyme, as well as those with low-activity enzyme, albeit at somewhat different environmental conditions.

In another form of balancing selection, that is, heterosis or heterozygote advantage, it is the heterozygote, that is, the individual with two different allelic forms of a specific gene, who is favored and selected for. One of the most widely known examples of heterozygote advantage is seen among those people who are heterozygote for sickle-cell anemia trait, a widespread blood disease. It has been shown that, although being anemic is a disadvantageous trait in typical environments, it is advantageous in areas with high incidence of malaria. This enables the people who are heterozygote for sickle-cell anemia to fight off malaria better than generally healthy, non-anemic people.

As expected, any evolutionary change brought about by the action of balancing selection is very gradual, and eventual fixation of only one allelic form, if it ever happens, could be either through random genetic drift, that is, by chance alone or due to a drastic change in environmental conditions favoring only one specific allelic form, that is, through directional selection. I should mention here that random genetic drift, or chance, as it is commonly known, is another evolutionary force that is indeed very effective in small and isolated natural populations, such as those encountered in small islands; or populations passing through genetic bottlenecks, in which after horrendous natural disasters, only a small fraction of the original population is expected to survive.

Stabilizing selection is completely a different type of natural selection, and is generally divided into two types, normalizing and canalizing selection (see next section, p. 141). Normalizing selection is the type of selection that Edward Blyth, a contemporary of Darwin, was advocating some two decades before Darwin published his theory of directional natural selection. This kind of selection does not cause any evolutionary change; rather, it works to preserve consistency within populations. It does so by eliminating extreme genetic variants not best adapted to a certain range of environments.

Normalizing selection seems to be a very common form of selection and probably operates more strongly during the early stages of organismal development. If so, it could be responsible for most cases involving miscarriage, which has a very high incidence in humans. It is also most probably responsible in eliminating deleterious mutations carried by individuals before they reach their reproductive age. An example of normalizing selection in humans is the birth weight of infants at about 7-8 lbs. Infants with lower or higher weights are shown to have lower survival rates. Although it is not known whether this trait is completely under genetic influences, a somehow similar trait —clutch size in birds—is well known to be determined genetically.

About two or three decades ago, a new evolutionary theory brought a lot of excitement to the genetic community. It was called punctuated equilibrium. It postulated that organismal evolution might not be through gradual changes as envisioned by Darwin, but rather organisms might be evolving by long episodes of preservation punctuated by sudden changes. This theory was based on extensive fossil records that showed ancient animals instead of showing gradual changes through millions of years, remained the same for a long period of time, and then suddenly showed new drastically different features.

So understanding the different types of stabilizing selection, the type of natural selection that preserves the consistency of life forms, should be as important as understanding the ones responsible for evolutionary changes. The action of normalizing selection through the elimination of extreme genetic variants somehow reminds me of how most human societies operate, in which too much traditionalism or too much liberalism will eventually limit the functions of its practitioners.

Hopefully, some recent findings at the DNA level (see the following section) may encourage us in refocusing our attention and trying to understand these long neglected but important forms of natural selection. Probably, it would be more helpful by changing its name from stabilizing/normalizing selection to something else, so people do not mix it with the principle of Darwinian/directional selection. Since it operates in the exact opposite direction to bring consistency rather than change. We could call it Blyth selection, after Edward Blyth, another Englishman, who described this type of selection in his articles written from 1835 to 1837, twenty years before Darwin proposed his theory of natural selection.

Before further discussing DNA conservation in more detail, I should mention here that this topic would probably be too technical for some readers, with no biology background, to follow. Therefore, if after reading the first few sentences of the following section, you feel kind of lost, it is better to leave this discussion behind and start reading the section after it called Our Evolutionary Past and the Origin of Modern Humans, on p. 149.

The Riddle of Evolutionary Conservation

I do not remember exactly when the phrase "evolutionary conserved" started to appear frequently in scientific literature, perhaps two to three decades ago with the advent of DNA sequencing technologies. What it means is the complete or partial conservation of DNA sequences of some genes across many species of plants and animals. Evolutionary conserved DNA is by no means a concept, and was originally observed by comparing the DNA sequences among many organisms. It was also observed by comparing the amino acid (the building block of proteins) sequences of many proteins and enzymes. So the evolutionary conserved phrase is equally used for proteins and enzymes (enzymes are specialized forms of protein that accelerate chemical reactions, and proteins are chemical structures constituting many parts of the cell; proteins are made according to the blue prints of their corresponding genes). Examples of conserved proteins are the many household enzymes responsible for everyday workings of cells, from protein synthesis to DNA replication. These two important cellular functions, along with many other cellular processes, are highly preserved among widely different species of plants and animals.

The existence of evolutionary conserved stretches of DNA/genes/proteins could only be rationalized through stabilizing selection because its function is basically to oppose any evolutionary change. Ironically, this point apparently has evaded the genetic community, which commonly treats this molecular phenomenon as a typical Darwinian selection. Indeed, the phrase "evolutionary conserved" sounds like a contrary phrase. After all, evolution is not in the business of conserving anything. Evolution means change, transformation in time from one form to another. The reason for coining such a contrary phrase is probably to reflect the point that the various animal or plant species that share these DNA sequences with each other are, evolutionary speaking, descended from the same common ancestor. Still, when choosing a name for a process such as "preservation of DNA sequences across many species," the emphasis should be on its conservatory nature and its maintenance through stabilizing selection.

The DNA sequence conservation, not only within genes, but also outside gene boundaries, has recently been shown to occur widely within various genomes. A small portion of these non-genic stretches of DNA is known to have regulatory functions. But about 97% of the human genome is made of these non-genic DNA sequences. Previously, they were known as junk DNA, now mostly called genomic deserts, since they are devoid of any functional genes. It should be emphasized that not all junk DNA, but rather a fraction of it, has been shown to be evolutionary conserved.

The recent widespread observation of conservation of some junk DNA sequences across many species has been very confusing to geneticists. According to Neo-Darwinian synthetic theory, these non-genic stretches of DNA, not having any obvious function, are not supposed to be under any selective pressure. Therefore, they should not be conserved at all, and were originally expected to accumulate random mutations, albeit very slowly, within the genome of any type of organism.

On the other hand, when genes make proteins with essential functions for everyday operations of the cells, they are said to be under selective pressure and natural selection (in this case, stabilizing selection) acts to preserve their DNA sequences. So whenever a random mutation occurs within one of these genes that leads to a drastic change in the function of its corresponding protein, the new protein becomes selectively disadvantageous and its carriers will be eliminated either during early developmental stages, or later before reaching the reproductive age. Consider, for example, DNA replication. Without consistency at the level of DNA replication, if too many changes are allowed to occur in replication machinery, organisms, whether simple or complex, will soon, after some limited generations, dwindle in their population size and may disappear rapidly.

Let's now return to the observation that even some junk DNA show conservation across many species. If they are under some sort of selection, such as stabilizing selection, then what do they do for the genome that has to be conserved? Actually, it is very hard to answer this question. So many different events are happening at the DNA and cellular levels that we currently do not have much understanding about them. For example, different chromosomes and the DNA within them are known to sabotage each other as if they are having chromosome wars.

In addition, our genome is littered with small pieces of junk DNA that jump around between different chromosomes called "transposable elements," trying to propagate themselves as much as possible. Some have called them selfish DNA, implying that they have no actual function for the cell, while others have proposed that they play important roles during embryonic development. Furthermore, a fraction of the junk DNA is made up of pseudogenes. This category of genes, while not predicted by Mendelian genetics, is a common feature of many genomes in which there are several copies of the same gene. Many of these copies, riddled with mutations, are presumably nonfunctional, and therefore, are called pseudogenes. But there is the possibility that some of them are actually functional during early embryonic stages.

Probably we could better understand the phenomenon of DNA conservation (whether genic or non-genic) across many species through another type of stabilizing selection, the so-called "canalizing" selection. This type of natural selection has not even been mentioned in most genetic books in the last three or four decades. However, it was widely advocated in the 1950s by another Englishman, C. H. Waddington. To describe his ideas about the canalization process, he frequently talked about organismal development as a perfect example of a canalized system. His rationale was that, regardless of different environmental conditions or genotypic differences, development would lead to generally similar results.

Therefore, canalization would act like a mask covering most unique genetic characteristics of each individual, and human development would become so canalized that all humans would end up quite similar to each other. Only major mutations and drastic environmental stresses would deflect the development from its regular course, leading to gross developmental abnormalities. A quite familiar case, signifying the need for considering the possible involvement of yet unknown developmental processes such as canalization, would be Down syndrome in humans. Those with this syndrome either have an extra chromosome number 21, a very small chromosome, or an extra piece of this chromosome attached to another chromosome. Occurrence of either of these two chromosomal abnormalities is always associated with mild mental retardation, as well as a host of several physiological maladies, eventually leading to a relatively short life span.

The most fascinating feature of this syndrome is that it bypasses all racial characteristics in such a way that those affected with this syndrome look quite similar to each other regardless of their original ethnicity. Presently, according to our current Mendelian genetics paradigm, there is no explanation for how such a profound feature could arise by one extra copy of some specific genes. On the other hand, these anomalies might be anticipated if viewed in terms of Waddington's canalization concept.

To better understand canalization, one should be reminded of the developmental constraints imposed on early embryonic stages. Early embryonic development has its own very strict rules on how to guide populations of migratory cells through seemingly endless steps leading to tissue differentiation and organ formation. The fateful whirlpool of these moving cells represents one of the most suppressed states that any organism has to pass through in its entire lifetime. To survive the rigid and directional process of the early embryonic development, Waddington suggested that canalizing selection would favor genotypes that would best provide a buffering against environmental shocks. This would mean showing limited responsiveness to the changes in the environment or to new gene mutations with small effects.

Since I took my first course in general embryology, when I was an undergraduate student at the University of Tehran, I have been fascinated endlessly with the early stages of development. During these stages, formless masses of cells migrate ever mysteriously in search of their final destinations, while at the same time, through induction processes, some of them would change the destinies of other cells. My second course, this time in human embryology was equally fascinating. The professor teaching these courses was herself quite interesting. I think she was about forty years old or maybe younger at the time. She was rather small and always wore high heels to make her taller.

She was the personification of respectful female scientists whom we are so accustomed to seeing in science fiction movies. Strangely enough, when I came to the US to study, since there were, and still are, very few female faculty members, I never had any chance to take any course with a female professor here. Unlike the other female professors that I took courses with at the University of Tehran, she would always come to the lecture hall very formally, dressed in a pressed white lab coat. But she was very stern. One stupid question and she would blast you with no mercy. Nevertheless, I could identify with her more than the other three female professors whom I took classes with. She was very knowledgeable in her field, and I still vividly remember her classes, and especially one of her statements. She used to say that there is no greater miracle than when a normal infant is born because, during the embryonic development, there are so many events in so many stages that could go wrong.

That is exactly why Waddington argued so adamantly for the possible role of canalizing selection during embryonic development to explain why fetal development as a whole is such a successful chain of events. It should be also mentioned that indeed many things do go wrong during the early stages of embryogenesis as evidenced by the high rate of spontaneous abortion (miscarriage) during the early months of pregnancies. Realistically though, it is very hard to give an accurate percentage of miscarriages because during the first month of pregnancy women are not even aware of their pregnancies.

Later, at Stanford University, I took another embryology course. This time a ten-unit extensive summer course about embryogenesis in marine organisms offered at Hopkins Marine Station adjacent to Monterey Bay. It was one of my best summers, going early mornings to nearby tide pools to gather live specimens, growing them in the lab in big tanks, and later, under a microscope, watching their little embryos passing through embryonic stages. I remember walking from the school very late every night uphill in thick fog, while being mesmerized by barely visible giant cypress trees slowly dancing in the air. I sure miss those days now.

There I got to know tunicates, commonly known as sea squirts, one of my favorite organisms. There are about 3,000 species of them worldwide. As adults, they are small pieces of dirty looking dark green stuff attaching themselves to coastal rocks, like little rubbery square boxes, not at all like a living being. But at the larvae stage, they are one of the most amazing creatures on earth. Seen under a microscope, they have a big head with a long tail complete with nerve cord and notochord. The notochord is a gelatinous rod that acts as a structural support for effective upward swimming.

I remember one day I was watching the larvae under the scope, and then suddenly the tail started to shrink right in front of my eyes. And soon, the whole tail was gone, and with it, was gone the opportunity to be a vertebrate as an adult. That is, an organism with backbone, instead of a shapeless mass of cells. It is postulated that millions of years ago, one of these little tadpoles (larvac), or somcthing like that, did not lose its tail due to some mutations and became the ancestor of all vertebrates, including humans.

What is astonishing about early embryonic stages is that all vertebrates follow more or less the same developmental patterns up to certain stages. Then they start to differentiate into different species. It is quite possible that DNA conservation across different species, whether genic or non-genic, is responsible for the preservation of these early stages. The most memorable episode of early embryonic stages, which is quite similar in different vertebrates, arguably is when the cells that later will give rise to eggs or sperm start their fateful migration from outside the developing embryo into its shapeless body.

They are called primordial germ cells and originally are located in a small cluster inside the yolk sac. Then during the fifth week (in humans), they start their long migration. These are very large cells, so are easily traceable. Their destination is gonadal ridges, primitive structures inside the body that later will form gonads, that is, ovary and testis in female and male, respectively. During their passage, many of these cells will vanquish, but some will reach their destinations. The cells derived from them, sperm or eggs, will carry the genetic material to the next generation.

It is proposed that the germ cells are set aside so early in the yolk sac in order to buffer them from many concurrent cellular divisions and DNA replications that other cells have to go through. The rationale behind this proposition is that the more rounds of DNA replication, the higher the chance for mistakes and occurrence of random mutations, and therefore, changes in the genetic material inside each cell. Sperm and egg cells derived from the germ cells are the only cells in the body that in essence are given the chance to become immortal. A few of them, after fertilization in a new host, go on to make a new embryo and the cycle goes on and on. So it is essential for them, the original germ cells, to be as much as possible, the exact copies of their original blueprints.

It should be mentioned here that in higher organisms in which sexual reproduction is the norm, transfer of the exact copy of genetic materials from one generation to the next is not possible. In each generation, parental DNA gets mixed (recombined) in germ cells after their migration into the body and before they produce egg and sperm cells in female and male, respectively. Why does parental DNA recombination happen? We really don't know. It has been proposed that it is a price that consistency of life has to pay to become more variable. New combinations of parental genes, resulting from the elaborate recombination process, are thought to bring selective advantage in case environmental conditions drastically change.

This notion, well accepted among geneticists, is not in line with one of the main dogmas of evolution which states "evolution has past, but no future," that is, everything is accidental and nothing is planned. Nobody really knows whether this dogma is accurate or just a fancy phrase. Nevertheless, there seems to be more to DNA recombination than just an "in case" scenario. The process of recombination called crossing over which occurs between two similar (homologue) chromosomes, one from each parent, is such a precise and ritualistic event that it is hard to imagine it has no immediate role in the subsequent early embryonic stages.

In summary, to fully understand our junk DNA, including the pseudogenes, we have to understand evolution and to do so, first we have to understand embryology, and to understand embryology we have to understand different types of stabilizing selection which operate in the opposite direction of Darwinian selection. The processes that separate living beings from still life are very complex and in essence confusing to us. Indeed, there are no easy answers. If we are in a rushing phase to find some answers to reflect our particular worldviews, whether shaped by our religious and traditional beliefs, or political and scientific agendas, that is one choice. But if we have no motives other than understanding our human origin, then there is no choice but to be patient and keep trying different ways of acquiring knowledge.

Our Evolutionary Past and the Origin of Modern Humans

Some years ago, I took a Super Shuttle from SFO (San Francisco Airport) to San Francisco. During the half hour drive, I and the shuttle driver started to talk about various issues and when he learned that I was a geneticist, he asked, "So what do you think about the human evolution? Do you think it had a single or multiple origin (or more accurately multi-regional origin)?" I was very surprised by his question, not only because he was aware of both theories, but also because he presented them on equal footings. Usually, in the popular press, the focus is on the "out of Africa" theory, that is, the single origin, probably because it sounds very dramatic and macho.

Currently, there are two competing evolutionary theories about how modern humans evolved into so many diverse looking ethnic groups. Before talking about these theories, one important point should be made. We humans, compared to many other animals, such as chimpanzees, show much less genetic variability, suggesting that humans share a much more recent common ancestor with each other than chimps do with themselves. It is an indication that, evolutionary speaking, humans are relatively a young and homogeneous species.

The assumption behind this notion is that, as time goes by, more DNA changes occur and accumulate within our genome. As a result, generally speaking, the DNA sequence of any given population becomes polymorphic (more variable) with time. In humans, DNA polymorphism, by and large, is quite low between different ethnic groups, which means that with all our real and assumed ethnic differences, genetically speaking, we are still quite a homogeneous species.

Before talking about the two theories of the origin of modern humans, first I should give a very brief summary about the origin of our ape-like ancestors as proposed by human paleontologists, the scientists who study the fossil bones of humans or pre-humans (hominids or hominins, as they are also called, is a reference to both groups). A more recent name for these scientists is paleoanthropologists, due to a recent trend in this field to figure out the path of human evolution by looking at fossil bones in connection with the stone tools left behind by the people to whom the bones belonged.

I would like to keep this section as short as possible because when dealing with a limited number of fragments of bones, vaguely estimated to be several million years old, this field of science becomes so murky and subjective that at times it looks more fictional than factual.

In the last 150 years or so, many fossil bones have been found in many regions of the world, especially in the Horn of Africa, that do not look like any bones of present-day humans or great apes. Nevertheless, they show varying degrees of similarity to bones of both humans and apes. So these long gone hominids have been proposed to be our common ancestors. In the past, typically these hominids used to be portrayed as a series of human-like creatures walking more and more upright, reflecting a single evolving lineage leading to the anatomically modern human. No more. Now that view is called inaccurate. The new image now is that many of those hominids represented evolutionary dead ends that were extinct millions of years ago and never contributed to our humanity. Only a few of them survived and became our direct ancestors.

What distinctly separates human fossils from those of the great apes are tooth morphology, different features of the skull, especially the size of the cranium (skullcap), and bones related to walking upright (bipedalism), such as the knees and pelvis. Since among the great apes, the behavior and genetic characteristics of chimps are closest to humans, it has been proposed that humans and chimps had the most recent common ancestors. One notable genetic difference between humans and the great apes, including chimps, is that they all have 48 chromosomes (two sets of 24 chromosomes, one set from each parent), whereas the human has only 46.

Chromosome banding patterns clearly indicate that the human chromosome number 2 has evolved by fusion of two primate chromosomes. It should be mentioned here that the human genome is made up of about 3 to 3.5 billion nucleotides, each a building block of DNA, and each prone to chemical changes resulting in one point mutation. But not all evolutionary changes are due to point mutations; some also involve the addition or deletion of a piece of DNA of variable sizes.

A rough draft of chimp DNA sequence that was aligned with the human one in 2005 showed that chimps and humans share 98.77% of their DNA sequences. At the same time, there are more differences between the human and chimp genomes than indicated by 1.23% base pair changes (a base pair is a reference to two nucleotides, each on a complementary DNA strand). Indeed a large amount of genetic material, about 2.7% of the genome, has additionally been inserted or deleted since an estimated 6 million years ago, when the chimp and human common ancestor split into two separate branches. Many of the insertions contain an unexpectedly large number of gene duplications in humans. Half of the DNA insertions and deletions in the human genome are specifically located near the tip of chromosomes (telomeres), rather than being randomly scattered all over the genome. Nobody currently has any idea about the functions, if any, of all these insertions and gene duplications in the human genome. In total, some 40 million evolutionary events mark the differences between the human and chimp genomes. However, the majority of these changes appear to be neutral, that is, to have no selective advantage. Some of the changes are seen in regions responsible for regulating gene activities. Major changes in gene activities could explain some of the major differences seen in the physical appearances of human and chimp.

As mentioned before, pseudogenes are non-identical copies of one particular gene, presumably due to gradual accumulation of point mutations. One of the more puzzling differences between the chimp and human genomes has been the rapid change in the human's olfactory (smelling) receptor genes. In particular, the number of pseudogenes has increased substantially. There are some 900 olfactory receptor genes in the human genome and some 70% of them are pseudogenes. It was previously proposed that since humans no longer rely strongly on their sense of smell to survive, these pseudogenes should indeed be regarded as genomic relics with no functions. Therefore, any changes (like any new mutation) in these genes, should be in effect neutral mutations. Nevertheless, many of the new changes could be under selective pressure, implying specific functions for these genes.

As suggested by some investigators, they may be involved in the human mating process. Since each individual seems to have a different set of the olfactory receptor pseudogenes, one possibility could be that they play a role in defining and displaying, through yet some unknown mechanisms, the genetic identity of each person. If so, an innate detection system involving rapid scanning of each other's genetic identity might be used in selecting proper mates with more genetic compatibility. In other words, these pseudogenes might be involved in determining the degree of attraction between two potential mates. As mentioned before, during early embryonic stages, there are so many events that have to take place exactly according to the genetic blueprint that a certain degree of genetic compatibility between egg and sperm might be required before it all gets started.

As indicated above, from the comparative genomic analysis of the DNA sequences in human and chimp, it has been proposed that their most recent common ancestor lived about 6 million years ago. So let's see, is there any fossil record supporting this proposition? The oldest fossil claimed to be a human ancestor is a partial skull of the upper face (without the lower jaw) and the full skullcap. It was unearthed in Chad, Africa, in 2002 and is called Toumai. Its age is estimated to be 7 million years old. However, not every paleontologist agrees that this skull belonged to a pre-human lineage; some assert that it is the skull of a female gorilla.

Another human ancestor, the famed partial skeleton named Lucy, who became a household name some three decades ago, is about 3 to 4 million years younger than the Toumai skull. Lucy was found near the Awash River in Ethiopia and was estimated to be 3 to 4 million years old. It was less than four feet tall and had a small skull like a chimp's, but its teeth were more human-like. Also, it had a bipedal knee structure with ape-like fingers. It did not have a full skull and its skeleton was not complete, either. But it was discovered with about 40% of its skeleton present, which is a rarity among fossil bones.

Its major contribution to the human paleontology field was discarding the commonly held previous assumption that the origin of humanity was due to an increase in its brain size. Here, it seems that walking upright was the turning point in the human evolution. Of course, due to the scarcity of fossil bones as old as Lucy, all these theories are highly speculative. But three decades ago, when the skeleton of Lucy was presented to the public, it was the heyday of public interest in evolution and many people were infatuated with the evolutionary scenarios presented to them by some enthusiastic scientists. I remember, around the same time, PBS was showing a very captivating TV program called *The Ascent of Man* about human evolution.

Coincidently, in those days, which was a decade after the start of the hippie movement in the US, few people showed much interest in Christianity or other monotheistic religions. Now times have changed. As a whole, people are gravitating more towards religious thoughts and are starting again to question what scientists tell them about the human evolution. Ironically, in the middle of all these events, people are learning more and more about the new technological advances in genetics and how it is going to transform our lives in the near future. So nowadays, it is indeed a very confusing time.

Now let's leave behind Lucy and the other very old prehumans, and look at much younger human fossils to find out more about the theories of the modern human origin, that is, how *Homo sapiens*, the wise human, spread to the four corners of the world and initiated the emergence of the human's different ethnicities.

The first theory is the single origin, or as it is commonly called, the "out of Africa" theory. According to this version of our evolutionary past, some 100,000 to 200,000 years ago, the modern human emerged as a new species in the Horn of Africa. Subsequently about 50,000 to 100,000 years ago, they began to venture out and spread across the world. It should be mentioned here that, by definition, species (as in a new species of modern-looking human, mentioned above) is a group of organisms capable of mating with each other, actually or potentially, and producing fertile offspring, but incapable of doing so with other such groups. At about the same time, there were Neanderthal people living in Europe and the Middle East, and another species of archaic humans, *Homo erectus*, were living in East Asia from China to Java, as evidenced by their bone fossil records and stone tools. Various human populations, when moving around, at least during the last 10,000 years, have always shown great tendency to interbreed with indigenous people, even with those they regarded as their slaves. It is just part of the human nature. But according to the single origin theory, somehow they never interbred with locals, and instead completely wiped them out and eventually replaced them, with no or very limited genetic exchange.

In multiple origins/multi-regional theory, humans migrated out of Africa in a much earlier time, about one million years ago. At that time, they did not look quite like modern humans, and their bones have been classified as belonging to *Homo erectus*, meaning "standing man." It is postulated that their evolution into modern humans, *Homo sapiens*, occurred independently in different continents. Since the modern humans from different continents have always been able to breed as one species, this theory entails that various populations inhabiting different continents should have enough contact with each other to ensure their unity as one interbreeding species. At the same time, they were separated enough to develop distinct physical features of their own as evident today in people inhabiting different areas of the world.

The multiple origins theory recently has gathered a lot of support especially among Chinese scientists. They strongly oppose the idea of the replacement of the local *Homo erectus* population of East Asia by the modern human from Africa, as envisioned by the single origin theory. Their objective is that the single origin theory does not take into account the existence of some specific physical features commonly shared only among present-day East Asians. One clear example would be the dental features of some East Asians, which are more similar to those of local *Homo erectus* fossil bones than to those of people of other continents. Another example would be the general flatness of face common to many East Asians. These examples clearly support the notion of continued evolution of *Homo erectus* of East Asia into the present-day modern human.

On the other hand, supporters of the single origin theory have raised their own concerns about multiple origins theory. They assert that when *Homo erectus* moved out of Africa some one million years ago, and established local populations in Asia and Europe, these populations could not have been large enough or widespread enough to ensure continuous contact between them, which was necessary to keep them as one single species capable of mating with each other, as all humans are capable of now. However, it should be added that inability to be in touch with each other, due to geographical distance, by itself would not necessarily lead to any reproductive isolation among different groups of people, resulting in the formation of new species. Reproductive isolation, first and foremost, is a function of the degree of genetic incompatibility between two potentially mating groups of people.

Genetic incompatibility, a measure of genetic differences, plays a crucial role in the first generation when, during egg and sperm production, the chromosomes of two different individuals first encounter each other for the process of pairing and genetic exchange. Then if there is not enough genetic compatibility between various chromosomes coming from each parent, eggs/sperm will not properly form, leading to sterility among the offspring in the first generation, and the subsequent reproduction isolation between the two groups of people.

Currently, there are at least two lines of scientific evidence that have relevance in evaluating the two competing theories about the origin of modern humans. We will first look at the human fossil records, then at some DNA studies addressing this issue.

Our Ancient Broken Skulls

Dating Methods Used in Estimating the Age of Fossil Records

Imagine that you are in the serene courtyard of the old Paul Getty Museum high above the cliffs in Malibu, California, overlooking the Pacific Ocean. Imagine that you are there to view a very unique and imaginary collection of all specimens of the human fossil bones unearthed during the last 150 years or so from all over the world. Looking at thousands of pieces, from single tooth to small fragments of broken old mandibles (jaw bones) and skullcaps cemented together to reconstruct ancient human skulls, it is easy to conclude that the human has evolved through time. But if one were assigned to collect relevant specimens to choose either the "out of Africa" origin of modern humans or the "multi-regional" origin, it would be almost impossible to decide between these two scenarios.

Right away one will notice that there are not many intact skulls or complete skeletons. There are a few, though not nearly enough to guide one through this huge task. However, if the intact fossil bones were not as scarce as they are now, still paleontology is a very ambiguous scientific field. This statement, by no means, is to diminish the efforts of many scientists who had spent their entire lives unearthing these relics hidden in harsh deserts. The ambiguity of fossil records is because the age of each newly unearthed specimen has to be estimated by some dating methods. Previously, a variety of relative dating methods were common, which are still in use in addition to several newer numerical methods. But the new methods have their own statistical and technical limitations as well.

The radiocarbon or carbon-14 is the most widely used isotopic technique in archeology, and as it is a very well known method, many people assume that it is the method of choice in determining the age of very old human bones. Indeed, it is not the case. Since carbon-14 has a relatively short half-life, which is 5,730 years, it can be reliably used only for specimens going back up to 30-40 thousand years which is, more or less, the range of the anatomically modern human. It should be mentioned that the half-life of a radioactive substance is a fixed time period required for the disintegration of half of its atoms.

To find out the origin of modern humans, our progenitor, the primitive human, this common method is of no use. Another isotopic method, radiopotassium, or potassium 40, is one of the current methods used for dating much older specimens. It works best on rocks, or rather minerals in rocks rich in potassium, while with radiocarbon method wood, charcoal and shells associated with fossil bones are dated. Although the radiopotassium method has no practical maximum limit, as radiocarbon has, nevertheless, it has another limitation. That is, with a half-life of 1.25 billion years, it cannot be used to date rocks younger than a few hundred thousand years old.

Therefore, when applying these two methods, the significance of the fossil records of the primitive human is limited by the difficulty of getting reliable dates between 40 to 200 thousand years. This is exactly the time range that would be of great help to trace our primitive human ancestry before it presumably started roaming around the globe.

Another commonly used dating method is paleomagnetism. This method relies on the observation that the earth's magnetic field shows frequent fluctuations in its intensity and direction. The shift in the direction seems to be due to reversal in the currents within the earth's fluid core. Each magnetic shift may persist as long as several hundred thousand years or as little as tens of thousands of years. Since the duration of several major and minor shifts in the earth's sediments has been determined for the last 5 million years, in principle it is possible to date specimens found within different layers of sediments. In practice, however, this method is less reliable than the radioactive methods.

Nevertheless, even if tomorrow a new and perfectly reliable dating method were to be introduced to the paleontology field, still the veil of ambiguity could not be completely lifted from this field. Another compounding problem is that when unearthing an old bone or stone tool, one cannot automatically assume that it had always been rested on the sediments it was found in. What if it was washed down from an upper layer, which is technically younger in age, to a lower and older layer of sediments? To counteract this problem, scientists routinely look for other nearby objects, either bones of other animals or stone tools. However, frequent earthquakes, floods and monsoon rains causing landslides have always been part of the earth's repertoire, capable of relocating large areas all at the same time.

Shifting Ideologies, Shifting Theories

Despite the scarcity of fossil records and the inherent difficulties in accurately dating them, many scientists in this field have come to agree on the following rosy picture of how, most probably, the primitive human had spread across the Old World. By about 1.8 million years ago, the descendants of Lucy (see p.153) in Africa had evolved into a hominid so closely similar to the human, as compared to apes, that indeed deserves to be called human. Her descendants, who are now called *Homo ergaster*, were taller and walked completely upright and possessed a relatively big brain, at least twice as big as today's chimp brain which is about 400 cubic centimeters. Soon, they started to make sophisticated stone tools, called Acheulean, replacing their less sophisticated tool making traditions, called Oldowan. The new tools, consisting mostly of hand axes, enabled them to venture out of Africa and spread all over Eurasia.

During the next 0.8 to 1.3 million years or so, they evolved into three geographically distinct human lineages which still showed a shared tendency for bigger brain size, smaller face and teeth. First to show noticeable differentiation were those who had spread into East Asia, and by one million years ago, they evolved into *Homo erectus*. Those who spread to Europe and western Asia were in closer contact with their African progenitors, and therefore, were evolved into a separate lineage much later, about half a million years ago. Their lineage is called *Homo neanderthalensis*, commonly known as Neanderthal. Those who remained in Africa were evolved into another lineage called *Homo sapiens*, our current progenitors.

By about 200,000 years ago, the brain size in all three distinct lines reached the same level as seen currently in modern humans (about 1,300 to 1,400 cubic centimeters), either through natural selection or random genetic drift, as postulated by this theory. Nevertheless, the morphology of the skullcap in the three groups became distinctly different. During this long period, the primitive human in all three lines showed little innovation in their stone tool making methods. In Africa, they were still making and using the large Acheulean stone tools including hand axcs. In Europe and western Asia, another tradition of tool making, called Mousterian, was later developed and became common. They were rather smaller in size than Acheulean tools. In East Asia, the less sophisticated Oldowan tools, consisting mainly of small choppers and scrapers, were still in use.

Then abruptly, about 40-50 thousand years ago, during the last glaciation period, Africa was swept by a wave of innovation in tool making, called Upper Paleolithic. They made narrow and pointed stone flakes and blades. They also started to shape bones and ivory, probably for hunting animals. The outcome of this newest innovation in tool making was not supposedly good news for the rest of the Old World. According to this theory, the people of *Homo sapiens* lineage marched out of Africa and not only spread all over the world, but rather wiped out Neanderthals in Europe and western Asia, as well as *Homo erectus* in East Asia, and replaced all of them in a matter of a few thousand years, with no interbreeding with them.

Ironically, not long ago, the birthplace of the ancestors of modern humans, the early *Homo sapiens*, was considered to be in the Middle East, based on relevant fossil records. Now, in the course of a few decades, it has suddenly shifted to Africa with no new breakthrough in fossil records. In addition, previously, scientists in this field used to express another point of view in which the early *Homo sapiens* supposedly hybridized with their contemporaries who had more primitive features, probably during the time when they first encountered each other through warfare.

One notable new finding in fossil records relates to the origin of Australian aboriginal people. Previously, their fossil records were estimated to be 30-40 thousand years old. But recently, using newer dating methods, they have been updated to be 50-60 thousand years old, which seems to precede the "out of Africa" expansion by 10-20 thousand years. Since Australian aboriginals look more like people of Africa than East Asians, it has been recently proposed that an earlier expansion from Africa to Asia might have taken a coastal route in addition to a later expansion inland. This coastal route presumably connected the Horn of Africa to southern Iran and India through Saudi Arabia and the Strait of Hormuz in the Persian Gulf, then via Southeast Asia reached the Indonesian Islands and then Australia. However, since Australia's fossil records of human bones are not nearly as old as some regions of East Asia, it seems the new immigrants had established new settlements in Australia, rather than replacing some previous ones.

Actually, the idea of an abrupt and complete replacement of two very old and established human lineages with another one, as advocated by the proponents of the "out of Africa" theory, is a very odd and hard to rationalize scientific view. It seems it was influenced not by discovery of new fossil records, but rather by the recent advances in DNA technology. Indeed, two recent DNA studies have drastically transformed the field of paleoanthropology. First, it was influenced by the results of a DNA study, as will be discussed in detail in the next section, comparing the mitochondrial DNA (mtDNA) of a group of living humans from across the globe.

Then, a decade later in 1997, another mtDNA study, this time carried out on a fossil bone, further supported the replacement theory. The most famous Neanderthal specimen supplied the fossil bone for this study. It was found in 1856 in Feldhofer Cave in the Neander Valley near Dusseldorf in Germany. The DNA sequence analysis on a small piece of mtDNA of this fossil bone showed 27 nucleotide (the building block of DNA) substitutions (mutations) when compared to a reference DNA. Subsequently, other studies analyzing the mtDNA from the bone tissue of three more Neanderthals showed more or less the same results. At the same time, the average number of nucleotide substitutions among living people, for the same fragment, was determined to be just 8. The results from these limited studies were trumpeted as the evidence that Neanderthals were genetically too different from modern humans, with no chance of interbreeding and genetic exchange between them. As noted by some other scientists, this conclusion suffers from a number of analytical shortcomings.

First, the rate of mtDNA mutation is not known, and currently it is assumed to be constant but it has been reported that it could be quite variable between different animal lineages. The second concern is that the human genome has some 3 to 3.5 billion nucleotides. Any result obtained from the comparison of just a few hundred of them is deemed inconclusive. It should be noted here that currently only mtDNA of fossil bones is amiable to DNA analysis. A very large percentage of DNA, called nuclear DNA, which is housed in the cell's nucleus, rapidly breaks down in old DNA samples, making them unsuitable for comparative DNA analysis. The third concern relates to the result of a very recent study that has seriously questioned the suitability of mtDNA for tracing human ancestry (see *Science* magazine, 28 April 2006, volume 312, p. 570).

As mentioned above, it seems that recently, for better or worse, the field of paleoanthropology has been greatly influenced by the recent advances in DNA technology. It is well known in the scientific community that scientific ideas are not always formed as objectively as commonly assumed. After all, scientists do not live in a vacuum and they could be easily influenced by the scientific trends of their times. Moreover, scientists do not live only in their scientific communities, but rather are part of larger societies usually infused with some particular political agendas. They, like any other member of any society, could be easily influenced by the subliminal messages continuously targeting humans in modern societies.

Of particular interest here are the consequences of a drastic change in the world's political landscape in the second half of the 20th Century. In the first half, the Western societies showcased the openness of their cultures to the traditionally reserved Eastern societies and concurrently empowered them with their advanced industries. Then, in the second half of the last century, the West turned around and embarked on a campaign to replace the ideals of Eastern cultures with those of Western cultures. The basic idea behind this replacement ideology is very simple: in the absence of diversity, everything will be more manageable to govern and thus profit from.

Now, apparently, the replacement ideology has found its way into the interpretation of DNA results and suddenly the Western scientists have come up with the notion that one group of early humans replaced all the inhabitants of the Old World in a big bang. So, with the gradual advancement of politics into science, it is easy to see that the resistance of some Eastern scientists to the "out of Africa" theory is as much politically oriented as the insistence of many Western scientists on the same subject.

Our Mischievous DNA

In Search of Mitochondrial Eve

The fascination of the general public with the "out of Africa" theory, one of the two major theories about the origin of modern humans (living humans), started about two decades ago, in 1987, with the introduction of the concept of "mitochondrial Eve" or "African Eve." It proposed that a single woman living in Africa some 200,000 years ago was our most resent common maternal ancestor; that is, we all descended from her. With such a name, considering the public's enthusiasm for learning about its distant past, combined with its naiveté about scientific affairs, it was just common sense to predict that the general press and thus, the general public would soon fall in love with such a concept.

Ironically, at the same time, many geneticists were wondering how such a small and intrinsically problematic DNA study could possibly arrive at such a profound conclusion. The study in question was involved in the sequencing of a few hundred nucleotides of mitochondrial DNA (mtDNA) obtained from some 150 people living across the world. I should mention here that the genome of every one of us contains some 3 to 3.5

billion nucleotides. Another similar study, a decade later, proposed that a man living in Africa some 60,000 years ago was our most recent common paternal ancestor, that is, we all descended from him. This conclusion was based on a DNA study using DNA markers on the Y-chromosome of 1,062 males from all over the world. So the name "Y-chromosome Adam" was given to this African forefather.

With such iconic names, the first thing that comes to mind is: "Why did these genetic Adam and Eve live so many thousand years apart from each other?" As is common knowledge, the popular press has a tendency to distort scientific findings to make them easier for the general public to understand. This is just another such case. The genetic Adam and Eve were meant to be hypothetical. Nevertheless, soon they were portrayed as if they were the Adam and Eve of the Old Testament. Furthermore, the idea of choosing these particular names was very tricky and misleading by itself.

Indeed, these hypothetical distant ancestors were never meant to live at the same time. More importantly, they were not the only two reproducing persons at the time. There were other reproducing males and females at the time who left behind many modern-looking descendants. But those other females, somewhere in the long line connecting them to us, left only sons and no daughters. Likewise, those other males, somewhere down the line, left no sons and only daughters. So the hypothetical Eve was the only female in her group who left an unbroken line of daughters. And the hypothetical Adam was the only male in his group who left an unbroken line of sons.

So the next question would be: "Why were our two hypothetical ancestors not supposed to live at the same time?" The explanation lies in the statistical methods used in creating phylogeny trees (such as family genealogy trees) connecting the living people whose DNA are used in such studies. The procedures for setting up phylogeny trees are such that they all invariably go back to a single common ancestor at some point in the past, and that is why we ended up with the hypothetical Eve in the mitochondrial DNA study, and the hypothetical Adam in the Y-chromosome study. Since these were two independent studies, there was no prior expectation that the single common ancestor in each study should be somehow related to the one in the other study.

The next step in making a phylogeny tree is estimating the date that the one single point at the bottom of tree represents, that is, when our most recent common ancestor in each study had lived. Each estimated date relies on two other variables. The first is a reference point that in each of these two cases represents the last common ancestor between the human and the chimp (estimated at 6-7 million years ago). The second variable is the mutation rate that is arbitrarily kept constant for all points in the line connecting the single point at the bottom to upper branches at the top of the tree. A further aim in constructing phylogeny trees for humans is to find out more about the possible migration routes that early humans took to spread across the globe. The many lines connecting the top branches of the tree to the single point at the bottom of the tree are used to explore possible migration routes.

The data points given in the phylogeny trees of these two studies each represent a unique DNA marker present in one or more people whose DNA was included in these studies. A human DNA marker is a small piece of our genome with known sequence that can be tested, in a DNA assay, to see whether it is present or absent in the DNA of the people under study. Although in a typical study a defined set of DNA markers are used, researchers working in this field are constantly looking for new DNA markers in people. Each new DNA marker is defined by a newly found mutation in the human genome, which is subsequently tested to see whether many people share it or whether it is a rare mutation, in each tested population.

In constructing phylogeny trees, the DNA markers that are shared with more people are set in the lower branches of the tree, implying that they represent more ancestral types. Nevertheless, as will be discussed later, there is an inherent limitation in the way these trees are constructed that makes the accuracy of the results of these studies questionable. Consequently, no matter how many DNA markers or individuals are tested in a study, neither the accuracy of the DNA sequences of the ancestral types nor the relationship between different ancestral types could be guaranteed.

Furthermore, it should be mentioned that when analyzing any data, the way these trees are constructed makes it possible to obtain not only one, but also several trees with different branching structures. Therefore, it will become a matter of personal choice to choose only one of them, which in essence could reflect the researcher's own particular scientific point of view. Before discussing very briefly the results of these two fateful DNA studies, it is important to state why mtDNA and Y-chromosome DNA, rather than nuclear autosomal DNA (see next page), are typically used in studies involving tracking the line of ancestry among several human populations.

Mitochondrion is a small organelle inside every cell of all plants and animals including humans. Its main function is to produce energy for the cell. There are about 1,000 of them in each cell, and each one has its own very small circular DNA making several enzymes and other proteins. One of its unique characteristics is that it is transferred only from the female parent to its offspring. The reason is that during sperm production, the small head of each sperm is filled with only chromosomal DNA materials. Although each sperm carries in its neck several mitochondria for energy production during its long journey to reach the egg, they are left behind when sperm is allowed to enter the egg. Thus, mitochondrial DNA is a good tool to track the line of ancestry from mothers to daughters in humans. On the other hand, the Y-chromosome is found only in the cells of male organisms due to the fact that males transfer their Y-chromosome only to their male offspring. So, Y-chromosome DNA is a good tool for tracking the line of ancestry from fathers to their sons.

Humans have 23 pairs of chromosomes in their typical body cells, called somatic cells, which are indeed diploid cells for having two sets of chromosomes, each set coming from one parent containing 23 individual chromosomes. On the other hand, only the egg and sperm, called gametes or germ cells, which are responsible for the transfer of genetic materials from one generation to the next, are haploid cells in the body, carrying only one set of 23 individual chromosomes. Each parent contributes one set of chromosomes to either egg or sperm. Twenty-two of these chromosomes have counterparts in each egg and sperm, and are called autosomal chromosomes.

The remaining chromosome is a sex chromosome, and is called either Y or X-chromosome respectively, depending on it being in a sperm or egg cell. Each female, in her somatic cells, has two X-chromosomes and each male, in his somatic cells, has one X-chromosome and one Y-chromosome. The X-chromosome is one of the largest chromosomes, while the Y-chromosome is one of the smallest ones in the cell. For this reason X- and Y-chromosomes cannot recombine with each other, while recombination (genetic exchange) between the other 22 pairs of chromosomes is a routine step during the process of generating gametes (egg or sperm cells).

One of the major reasons that the Y-chromosome is an effective tool for genetic identification is because it does not participate in chromosomal/DNA recombination. Recombination in genetics makes everything much more complicated. That is why geneticists cannot rely on autosomal chromosomes to establish lines of ancestry. This fact constitutes a severe limitation in our ability to reach an accurate estimation of when and where our ancestors started to move around the globe.

As some may remember, a few years ago, a case involving Y-chromosome DNA for genetic identification made front-page news. A person of African-American ancestry claimed to be a descendent of Thomas Jefferson, the third president of the US and a white American, who died in 1826. The comparison of some stretches of Y-chromosome DNA from Jefferson (as left behind in his hair strands) and the claimant clearly indicated their relatedness.

Some years ago, there was another famous case in which some Ethiopian Jewish people with dark skin claimed that they were the descendents of reunion between the legendary Queen Shiba and King Solomon, a legendary ancient Jewish king. Although DNA tests are not in the business of verifying legendary figures, comparison of the Y-chromosome DNA between Ethiopian Jewish males and living Israeli males clearly showed their genetic similarity, at least with respect to the Y-chromosome.

A second and equally important reason for using DNA from the Y-chromosome and mitochondria as tools for genetic identification is that these stretches of DNA have been assumed to be under no selection pressure. The basis for this assumption is that these DNA have accumulated many more mutations than the DNA of autosomal chromosomes.

Random mutations always occur in the DNA of all organisms, albeit with very low frequencies. If a piece of DNA, such as a gene, has an essential function for the cell, its sequence tends to be more preserved, as shown through comparative analysis with other closely related species. The current rationale for using the Y-chromosome and mitochondria as tools for genetic identification is that if a gene or a stretch of DNA is not under selection pressure, then it is free to accumulate random DNA mutation that naturally occurs in every organism. If so, then the rate of mutation accumulation becomes a function of time. The more differences between two stretches of DNA, the more time has elapsed since they separated from a common ancestral DNA. This concept has been compared to the function of a clock, ticking at exact intervals, and has been used to introduce theoretical molecular clocks in genetics. These clocks have been widely used for genetic chronology employing both DNA and proteins.

As mentioned above, DNA markers on the Y-chromosome have successfully been used to relate Jefferson, who lived some eight generations ago, to his living descendents; or living male Ethiopian Jews to some living male Israelis. But there is a distinct difference in these cases and the effort to track down a line of ancestry going back some 2,000 to 3,000 generations to find out more about the origin of modern humans.

There is a major intrinsic problem involving the concept of the molecular clock in genetics. It assumes that the rate of mutation is constant, regardless of the past or present environmental conditions of the organism. Just because a piece of DNA now behaves as not being under selection pressure does not mean that it has always acted so. There is also no basis for assuming that mutations always occur randomly within the genome. For example, some DNA regions, in principle, could be more prone to DNA replication errors leading to new mutations.

To have a better estimate of the mutation rate, which was considered constant in the mtDNA study, the Y-chromosome study used another approach. First, in this study, single nucleotide changes in the DNA sequences of specific markers were tracked down; then, the changes in the microsatellites in the vicinity of the single nucleotide changes were also considered. Microsatellites are repetitive DNA sequences of nucleotides that are repeated usually in tandem like CACACACA, which is four repeats of CA or CAAGCAAGCAAG, which is three repeats of CAAG (C, A, and G here each represent a different chemical base, which is part of the nucleotide structure). As predicted, there could be numerous variations of these tandem repeats. These repeats are most often in the non-genic junk DNA regions of chromosomes, and the Y-chromosome is particularly littered with them.

The reason for using microsatellites is because they show a much higher rate of mutation. Due to their repetitive nature, there is a much higher chance of error being made during their DNA replication (copying); that is, the number of repeats may increase or decrease by one unit at a time. In the original Y-chromosome study, it was assumed that the rate of mutation during DNA replication for all microsatellites was the same. But later it has been shown that each different type of repeat has its own rate of mutation. This new feature will further complicate the accurate estimation of genetic chronology.

Going back to the subject of mutation in regular DNA sequences, in the synthetic theory of evolution, mutation itself (regardless of its rate), by default, was defined as a spontaneous random event. A series of studies in the 1950s were especially designed to validate this assumption. But later, several studies involving directional mutations (not random) have strongly challenged this view. If further studies demonstrate that indeed some mutations are directional, then the concept of the molecular clock no longer could be considered as valid.

"Least Effort" Genetic Findings

Phylogeny trees constructed from mtDNA and Y-chromosome DNA also ignore one important issue when considering mutations. That is, some mutations could occur at the exact same position more than once, without any chance of being accounted for as more than one mutation. We encounter this problem because for the sake of simplicity, the current method of choice in constructing phylogeny trees applies the concept of "minimum evolution." This concept refers to a philosophical argument known as parsimony, commonly applied in the West but frowned upon in the East. Based on this view, nature works according to the "least effort" principle. Let's see how it has been applied to DNA mutation studies. It should be mentioned that the DNA building blocks are made of four different nucleotides (specific chemical compounds). For the sake of brevity, their abbreviations, A, C, G, T are commonly used.

Let's suppose that there are two very small pieces of mitochondrial DNA, each from a different person, that we want to compare with each other to assess their common ancestry: ACGTTGCA and CCGTTGCA. We notice only one difference at the position of the first nucleotide, that is, A and C. Now assume that in the DNA of the most recent ancestor, A was also at the first nucleotide position. So it could be assumed that some time in the past in the line of the ancestors leading to the second person, A had mutated to C.

Furthermore, there is a slight chance that in the line of the second person, first A muted to G, for example, and then, at a later time, G mutated to C, its current structure. The end result is that two mutations had actually occurred in the line leading to the second person, but presently we could notice only one. Likewise, if a small piece of mitochondrial DNA at a specific region is exactly the same in two persons, it is not necessarily an indication that they shared a common ancestor in the recent past.

Another concern in constructing phylogeny trees is how to choose a relevant point of reference. Comparing two pieces of DNA sequences with each other cannot in itself establish any genetic chronology. Let's assume that there are 5 base pair differences (the same as 5 point mutations) between two pieces of mtDNA belonging to two individuals from two different populations. Simply by looking at these differences, one cannot establish which piece more likely represented the ancestral DNA.

If later the same piece of DNA, from a third individual belonging to another population, is found to have 10 base pair differences, then it is said that the third piece is ancestral to the other two DNA. Thus, the relative ages of the first two DNA are decided by comparing them to the third DNA. The third piece of DNA is said to be ancestral because, as mentioned before, having more base pair changes is assumed to mean that it had more time to accumulate more mutations.

This basic assumption, however, is based on other assumptions, mainly as discussed before, that mtDNA is under no selection pressure and thus, the mutation rate would be considered constant among all populations. But the last two assumptions are not empirically based, but rather are inferred from other observations.

Now let's go back to our two phylogeny trees that were constructed based on the mtDNA and Y-chromosome DNA studies. In both cases, each tree had two major branches. One branch consisted of the DNA markers of African people, and the other branch consisted of those of some Africans and all other people in the study. Since the people in the first branch showed more variation in their DNA (in this case point mutations) than those in the second branch, it was presumed that the people in the first branch represent an older evolutionary line and thus, the origin of modern humans must be from Africa. But at the same time there are other phylogeny trees that have been constructed using the same data that show different branching that do not support an African origin.

Another conclusion from these studies is that the ancestors of the African people (living in the north and northeast) who do not share any DNA markers (used in these two studies) with other Africans were the hypothetical "Adam and Eve" who migrated out of Africa. But equally, it could be assumed that these very same people migrated into Africa from somewhere else such as Eurasia. And that is why their descendants do not share specific DNA markers, tested in these studies, with other Africans (see the results of a recent mtDNA study making the same conclusion, published in *science* magazine, 15 December 2006, volume 314, p.1767).

Regardless of what phylogeny trees tell us about the origin of modern humans, whether it is from Africa or we indeed have multi-regional origin, what the mtDNA and Y-chromosome DNA studies both unambiguously indicate is that genetic variability/diversity among humans is much higher in sub-Saharan Africa. There are two ways to explain this outcome.

First, there is the possibility that non-African populations had recently more contact with each other and thus exchanged more genes, resulting in a decrease in their genetic diversity. The vast Sahara in northern Africa, which geologically speaking is a recent formation, may have contributed to the genetic isolation of sub-Saharan Africans. The second possibility is that non-African populations all share a more recent common ancestor with each other than they do with Africans.

Let's now see, generally speaking, what factors could contribute to the genetic diversity in any given population:

Time: as discussed above, more variability could be an indication of longer evolving time.

Population size: as the population size increases, there is going to be a higher chance of genetic diversity.

Mutation rate: higher Mutation rate means higher genotypic diversity, which may or may not translate into higher phenotypic diversity. With no empirical data currently available, mutation rate routinely is considered constant for all different environments. But there is the possibility that it could be higher in environments with higher/lower temperatures.

Type of selective pressure: balanced selection tends to increase polymorphism (variability) in a population. Balanced polymorphism is usually preserved in a population when each allelic form of a gene is more suitable for a special niche of the environment. For instance, many environmental factors that contribute to endemic diseases could help sustain balance polymorphism in a population.

Degree of inbreeding: some tribes are very restrictive and actively enforce within-tribe marriage, while other tribes may be less restrictive, which results in more genetic variability in the population.

Mating practices: it is well known that assortative mating can affect the genetic structure of a population. Sometimes people sharing specific traits tend to mate with each other more often. There is also the possibility that monogamy reduces the genetic variability of a population. Consider, for example, six siblings all sharing the same mother and father versus six siblings being half-brothers and sisters by sharing either two fathers, or two mothers, or two mothers and two fathers at the same time. Typically, the ones with the same mother and father look alike more and so are genetically more similar to each other.

Due to the special mode of inheritance of Mitochondria and the Y-chromosome, not all these factors are applicable to them, as they are to nuclear DNA. Nevertheless, since some of them, individually or together, are capable of operating within any population to affect its genetic structure, it is unrealistic and premature to assume that the higher genetic diversity of the African people is just a function of time.

It should be mentioned that the higher genetic diversity of the African people has been shown not only with the mtDNA and Y-chromosome DNA, but also with other types of DNA. Furthermore, before the current practice of using DNA markers, gene frequency changes across vast geographical regions were widely used to establish relatedness between different human populations. These studies have also indicated higher genetic diversity among Africans. But as mentioned above, there are many factors that could contribute to genetic diversity among various populations.

I started the section "Evolutionary Forces and the Myth of Darwinian Selection" with a discussion about the science writers of newspapers, and non-evolutionary geneticists, in general, not really paying much attention either to the big picture or to little details about scientific concepts relating to human evolution. As I wrote, one day they talk about human evolution through Darwinian selection and "survival of the fittest," and another day they talk about the African Eve, venturing out of Africa some 50,000 years ago, and in essence laying the genetic foundation of our modern humanity; as if they are really talking about the same underlying evolutionary force. Of course, science writers, generally speaking, are not scientists, and unfortunately most times, it seems they are out there just to entertain their naïve readers, rather than engaging them intellectually.

However, when some scientists assert the validity of the "out of Africa" theory and thus, the notion that the higher diversity of Africans is just a function of time, it is indeed another way of saying that human evolution was not driven by different forms of natural selection after all. In other words, it was just the consequence of a slow mutational force shaped by chance events, that is, random genetic drift.

The most serious issue with regard to the "out of Africa" theory, however, is not whether it supports random genetic drift or natural selection as the driving force of human evolution. The problem is rather technical, that is, it is founded on very limited DNA data based heavily on certain assumptions. The human genome is made up of the nuclear DNA, spreading over 23 pairs of chromosomes, plus a very small mtDNA located outside the nucleus. Until more data from the nuclear DNA becomes available, we are not even in a position to start exploring the subject of the origin of modern humans at the DNA level.

Furthermore, I started the section "Our Evolutionary Past and the Origin of Modern Humans" by recalling a conversation between myself and a Supper Shuttle driver many years ago, in which he asked me whether I thought that anatomically modern humans all originated from Africa or had multiple origins based on other continents. The answer I gave him then is the same one I have now. I told him, "There are so many unresolved, more essential issues in evolutionary studies that finding the answer to this question is not, and should not, be the focal point of evolutionary studies."

But when recently I went online and checked Wikipedia to see what kind of evolutionary topics are discussed there, I was amazed to see how much this topic has attracted people. But surprisingly, I found that most entries were talking about the "out of Africa" theory of single origin. I suppose it is much easier for the public, particularly people with a limited evolutionary background, to make sense of this theory.

In summary, it seems that the application of current DNA markers for defining our migratory routes during our long evolutionary past is based on a set of assumptions. As someone with extensive working experience with these DNA markers, as tools for human identification applications, I am quite familiar with their limitations, as well as their usefulness. They are powerful only in defining our recent past; otherwise, they merely represent tentative answers waiting for other lines of evidence to support them.

We should remember that reconstruction of our evolutionary past is an immensely daunting task for geneticists. With ever advancing DNA technologies, we could sequence the entire human genome of thousands of people, and still not have an accurate account of what really happened to us in the past. One reason is that our current tools used to dig out our evolutionary past are based on assessing the mutations that differentiate us from each other at the present time. But there were many mutations in our genome in the past that had been lost with no footprints left behind. Those mutations, as well as the current mutations, are equally essential for an accurate reconstruction of our evolutionary past. Unless we come up with a new reconstruction method, looking into our DNA history will be more entertaining than informing.

Our Recent Evolutionary and Non-Evolutionary Changes

According to Stanford anthropologist Richard Klein, something extraordinary happened some 50,000 years ago to the human brain that suddenly made humans much more inventive in their very traditional tool making methods and also their way of life. Ironically, he is not talking about the brain size because, generally speaking, during the last 100 to 200 thousand years, the human brain size has not increased, most probably due to reaching the physical constraint of the human female pelvic size.

Based on fossil records, back then humans showed some distinct forms of skullcaps, but they had more or less the same brain size. Even the Neanderthal's brain was as large or even larger than the brain of people living today. So the brain size of modern humans by itself cannot explain human's sudden advancement in tool making, artistic expressions, or social organization. In other words, in our recent evolution, the brain size is not directly correlated with the increase in our intelligence. Klein is not talking about the emergence of some specific linguistic abilities that previously were assumed to be the basis for further development of human's superb intellect, either. He is vaguely talking about a mutation(s) responsible for a drastic neurological change in the human brain's organization or its circuitry.

His proposal is based on the observation that over a million years, humans showed little departure from their routine traditions of tool making, and then about 50,000 years ago, they started to show great innovation and risk taking in whatever they did. Nobody has any idea whether he is onto something or whether this is pure speculation; and even if his idea is correct, whether the neurological change he is talking about was adaptive or arose by genetic drift. In the last chapter of this book "Who Might We Be?" I will propose that this change might not even have been an evolutionary event, just a one-time biological event in the recent history of humans. Indeed, it would be a fundamentally different type of biological change as we currently define it for humans, although we routinely ascribe this kind of change for lower organisms, such as bacteria.

Whatever the nature of this plausible neurological change, there are other intriguing observations relating to human evolution. For example, two years ago, scientists, while analyzing 750 fossil teeth dating back many thousands of years, reported that around 30,000 years ago there was a drastic jump in humans' longevity. Again, we do not have any understanding about what caused this fundamental change in human biology.

Another recent change in humans, during the last 10,000 years, has been a major trend towards rounder heads and smaller faces and jaws, factors contributing to more refined features in our skulls. Ironically, many scientists agree that these changes might not be adaptive at all or even evolutionary changes. It seems that these changes are mostly due to how we use our jaws during the chewing process.

After the agricultural revolution some 10,000 years ago, and the invention of cooking methods, humans started to eat much softer food resulting in a considerable decrease in the chewing duration, thus relaxing the stress on the face and jaws. Consequently, the chewing muscles may have grown smaller, causing a similar reduction in the bone growth. The recent decrease in the size of the human jaw is reflected in a concurrent worldwide increase in tooth malocclusion, tooth crowding and impacted molars, outcomes with no adaptive values. As nowadays we automatically assume that everything has a genetic basis, there is an expectation that a new mutation has to explain every little biological divergence. However, some biological changes actually reflect the innate plasticity endowed in the living process, in this case the great plasticity of bones and muscles.

Another noticeable biological change not due to adaptation and natural selection seems to be an increase in human height. It is thought that an increase in the average height as seen in many developed countries during the last 150 years is mainly due to better diets; although, at the same time, the effect of chlorinated water resulting in less debilitating diseases during the first few years of childhood cannot be completely ruled out.

Our Transient Ethnicity

Probably, in less than 5,000 years, there won't be any Iranians around, or any Americans; not just by name, but also in the way we look, talk and believe. There will be other people, however, looking somewhat different, talking in a different way, and believing differently; and still occupying the same geographical locations we now call Iran and the USA, albeit with different boundaries. Some may question the credibility of this futuristic genetic scenario. After all, it is well known that evolutionary changes occur very slowly. Indeed, some evolutionary forces such as natural selection act very slowly in changing the genetic make up of human populations, in the replacement of old characteristics with the new ones.

The effect of natural selection has even been slower during the last 10,000 years of human history. Since then, we humans have lived in vastly different geographical regions, with different environmental conditions, which in theory could have enforced different selective pressures on us. However, in reality, constant technological advances during this long period of time have standardized the living conditions of humans across the globe to a point that both currently and in the foreseeable future, there won't be much selective pressure against genetically disadvantage people. Therefore, these people would be able to reproduce and contribute to the next generation, more or less, as many offspring as those endowed with more beneficial mutations best suited for a particular environment. In other words, if natural selection were the only evolutionary force, not many noticeable evolutionary changes would be expected at population level in 5,000 years, implying that, in the future, different ethnic groups would look more or less the same as they do now.

But as discussed before, natural selection is not the only evolutionary force and in its absence another evolutionary force, that is, gene flow, will be the main force to shape our evolutionary path in the future as it has done in the past. Gene flow, which is the natural outcome of human migration and its consequential genetic admixture between newcomers and indigenous populations, is indeed a very fast and effective evolutionary force. In the future, as is happening now, mating and genetic exchange between different ethnic groups, more than ever, will help homogenize the gene pool of the human species and significantly change the present-day characteristics of different ethnic groups.

Another very effective evolutionary force, random genetic drift, which, just by pure chance, would cause the fixation of many genes in small populations, could have a limited evolutionary impact in the future, similar to natural selection. Of course, it will be completely another scenario if in the near future horrendous natural or man-made disasters exceedingly reduce the number of reproducing humans to the point that only a limited number of people will be able to contribute offspring to the future generations. Random genetic drift works effectively in small human populations, such as tribes with strict marriage arrangements, making them in effect similar to small inbreeding groups, like those living on an island. Currently and in the future it is expected that tribal influences in marriage practices would be greatly diminished.

Many physical features of humans across different ethnic groups, are believed to be not the consequences of natural selection directly, but of random genetic drift in the past, when humans used to live in small groups, for example, in tribes. It also most probably shaped human evolution after recurrent natural disasters, such as earthquake and famine, or man-made disasters, such as ravaging wars; all drastically reducing human population size, at different places and different times. Any mutation, whether beneficial or harmful or even neutral, in the remaining small group of people, would have a great chance to get fixed in a few generations and then propagate widely as the small population would grow in size.

Going back to our futuristic scenario, it is common knowledge that the US is made up of many different ethnic groups due to the constant immigration of various people to this country during the last 500 years or so. It should be mentioned that, presently, the people of the United States, culturally and physically, do not really represent the proverbial phrase "melting pot," rather they are more like a "pot of soup" with distinct ethnic ingredients living side by side. However, there is more to this "pot of soup" than what meets the eye. For example, when comparing the genetic makeup of African Americans with that of the people of sub-Saharan Africa, where they originally came from a few centuries ago, there is a big surprise waiting even for geneticists. Indeed, it is estimated scientifically that, on average, the genome of each African American is 30 percent mixed with that of white Americans.

Furthermore, as a consequence of more recent rapid cultural changes favoring liberal attitudes, such as interracial marriages, there is the prospect that the current ethnic groups in the US will go through genetic admixture much more readily in the future. If so, there is a great chance that in the next 5,000 years the population structure and genetic makeup of the people living in the US will go through drastic changes, bringing new physical characteristics to its inhabitants.

On the other hand, most people living outside the Middle East assume that Iranians, like other people of this region, all look alike. In fact, nothing can be further from reality. I still remember the puzzled look of my Japanese friend, when last year she came to an Iranian gathering. Later she grumbled, "In Japan, we more or less all look alike, but how come here some Iranians look like Europeans, some like Arabs, and still others like Indians?" And that is not all. Apparently, some 75 different languages and dialects are currently spoken in Iran. These languages are remnants of Iran's past, when 41 nations across 3 million square miles constituted the Persian Empire some 2,500 years ago. So there are also great opportunities for various Iranian ethnic groups to mix genetically and look distinctly different 5,000 years from now. Then, imagine the genetic admixture of thousands of various ethnic groups all across the world during the next 5,000 years. Undoubtedly humanity will look very different by then.

If indeed our genetic identity as an ethnic group is so transient, then why should our ethnic identities matter to us so much? Why should I, while living in San Francisco and not having visited home for over 30 years, still cling to my Iranian identity like a little kid hanging to her mother's skirt? Why I am trying to explain my Iranian heritage to a world getting more and more comfortable in a standardized way of thinking and absorbing everyday events? I suppose I am doing this for the same reason that I left my well-paying job in a world-class biotech company, working in a nice cubicle with large windows overlooking the San Francisco Bay. I left that job and ended my 30 years of experience in science because I found it too frustrating to be constantly discriminated against, not so much because I was a non-European foreigner, but rather because I was a research scientist. Big companies have developed a corporate culture that idolizes the marketing people, deeply annoying and frustrating the research staff. Most research scientists, however, act wisely and get used to daily discriminations, knowing that in the long run they will be better off both professionally and financially. But for better or worse, that kind of wisdom will never be my guiding light. I left my job exactly because I did not want to get used to discrimination. It was completely a new experience for me, and as a geneticist I felt very disturbed by it. Genetics may seem to be foremost about our unique individuality, but at the same time, it is about the realization that we humans all have the same genetic blueprint. Even a more remarkable reality is that we are all the same inside, beneath our fancy skin dress that comes in so many shapes and colors.

Just looking at our facial features, such as our noses, lips, upper and lower eyelids, even our eyebrows, obviously these are all distinguishably different in different ethnic groups. Our bones are different too. Considering the human skull, the shape of the forehead, cheekbones, chin, brow ridge, lower jaw, the area above our lips, all are noticeably different, both at individual and population levels. Each of these features, taken by itself, could be seen in any person, regardless of ethnic identity. But when specific forms of several of these features always bundle up together, then this is what makes me an Iranian, and my next-door neighbor a Chinese.

However, our organs inside us have a different evolutionary history. Have you ever extracted DNA from the brains of dead old women, late at night in a quiet lab, all by yourself? If you have, then you know that our insides are a completely different story. Indeed, the results of a very recent study have indicated that the human brain size is evolutionary more stable than that of any other organ inside the human body. According to this study, the brain size is even more genetically influenced than height, and it is seemingly under no environmental influences. Also, there is no such a thing as a Turkish heart or an Arab liver. They all look the same. Beautiful, expressive eyes have nothing to do with eyeballs. All eyeballs are just the same; only the details of the upper and lower eyelids are different. Another step further down, at the cellular level and then at the organelle level inside our cells, everything has the same shape again. All the molecular machinery that forms us, makes our bones, muscles, nerves, blood components, hormones and enzymes, and maintains us is virtually the same.

While our molecular machinery has been evolutionary preserved to function the same way, probably through normalizing and canalization selections (see p.136), as we are slightly different at the DNA level, what they make or the amount they make could be slightly different from one person to the next. Our DNA is made up of some 3 to 3.5 billion nucleotides, which are the building blocks of DNA, and on average we are different from each other one nucleotide per 1,000 nucleotides, or about 3 million nucleotides. But remember that about 97% of these are located in non-genic regions of our genome, where no genes are located.

Therefore, structurally speaking, when looking at our flesh and bones, we can see that we are remarkably similar. This figure of one nucleotide difference in every 1,000 nucleotides is much lower than is the case for many other animals. So, humans, as a species among other animal species, are more genetically similar, which could be an indication that we are relatively a new species, having gone through less time to diverge significantly from each other.

But whether we like it or not, culturally we are not as close to each other as we are genetically, or as we would like to be. There remains a big divide between the cultures of the Old and the New Worlds. There is a saying in the East that asserts, "In the East everything matters, nothing works; in the West everything works, nothing matters." As I mentioned in the beginning of this chapter, about three years ago, I was so bored with American TV programming that I switched to an international TV channel. But I am sure that these days, many people would agree that American movies and TV programs have indeed become very boring. In reality, they are not the ones that have become boring; rather it is we that have become so dull. The media are just showing us the way we are.

I read somewhere that in the future humans will become pets of robots. I am afraid that our future might be even bleaker than this cute little forecast. I am afraid that, more likely, we will become robots ourselves. Progress through principles of advanced technology and single-minded mentality has idealized for us a robotic-like nature. Highest efficiency in whatever we do has become our gold standard.

But this new and regimented life style is not inspired by a Zen-like disciplined way of life, in return for a chance of attaining contentment and peace. We do it just because we see others are doing it. Just like a bunch of robots. Apparently, the quest for programmed perfection has made us increasingly boring and dull. That is the price the Western culture, our current dominant living norm, has managed to negotiate for us. But we still remember that the real gold never tarnishes, and sincerity never becomes dull.

Who Were Neanderthals?

I was going to see an Iranian movie at the Opera Plaza Cinemas on Van Ness Avenue in downtown San Francisco. It was called *Turtles Can Fly*, a movie about the plight of some Kurdish children in northeast Iraq, close to the Iranian border, just before the start of American war in Iraq in March 2003. My sister, who had already seen it, warned me that it is an extremely sad movie. It was about a group of orphaned kids who were teaching each other how to locate mines and detonate them and then sell them in a nearby market for a small amount of money, as a means of survival in that rugged mountainous area with no parents left to take care of them.

The little kids were representing the youngest victims in a new venture by the invading forces of the West, bringing chaos to an ancient land. The world-shaping zeal of American mentality was marching into the Middle East, yet another time, and crashing head on with the traditionalism of the Middle Eastern people. I wonder if these two distinctive cultures were like two organisms being subjected to evolutionary forces, what kind of creatures might they have represented? One is akin to a multifaceted and colorful, but out of proportion, monster, afflicted with a high rate of mutation and constant change, and the other, more like a dinosaur barely evolving through time. So, evolutionary speaking, neither of them seems to have the capacity to survive in the long term.

While I was walking in the Civic Center towards Van Ness Avenue, to see the Iranian movie, I noticed several floats starting to move along one of the nearby streets. On one of them, there were several performers, standing in two orderly rows, banging on their big drums and making an awesome sound. For a few hours, I totally forgot that I was there to see a sad movie. It was a beautiful spring afternoon and the people standing along the parade route seemed very excited to see many different performers in their traditional dresses playing traditional roles, accompanied with loud music echoing everywhere.

How ironic, on one hand we cheer people's traditionalism on stage, and on the other, we cannot quite understand why some people want to retain their traditions in real life. Or maybe it is not as much about disliking traditionalism in others as it is about being unable to control our primitive instincts, such as fighting and grabbing each other's possessions. Ironically, it seems that setting up big armies of people and spending much time and effort to bring new innovations to our warfare arsenals, for the sake of killing each other over limited resources, is in itself a prehistoric tradition we modern humans cannot easily discard.

Nevertheless, many people, including scientists, are still adamant in denying the possibility that the primitive fossil bones scattered widely in Europe and western Asia and belonging to creatures known as Neanderthals were related to our ancestors. Neanderthals, like dinosaurs, have captivated the public attention so much that since their discovery in the 19th century, they have created their own myths, far removed from the actual scientific evidence. So who were Neanderthals anyways, the creatures we have been so fascinated with and at the same time so afraid to be related to?

The movie *Turtles Can Fly* that I talked about in the beginning of this chapter was filmed in a Kurdish village located in a vast triangular region shared between Iraq, Iran and Turkey. It is the ancient homeland of over 20 million Kurdish people, a proud and traditional people, living in the highlands of the Zagros Mountains zigzagging the Iran-Iraq border.

Some 10,000 years ago, cultivation of wheat in these fertile valleys brought on an agricultural revolution that soon fostered the development of an ancient civilization in the Middle East. Kurdish people, who speak the Kurdish language (linguistically a member of the Iranian family of languages), are generally distinguished from other people in the Middle East by their lighter skin, eye and hair colors. They are well known for their beautiful traditional music, dance and colorful costumes, especially the baggy multi-pleaded and well-designed trousers used by men.

Somewhere in the same region where this movie was filmed, there is a well-known cave called Shanidar, in the shadow of the 12,000-foot Haji Ibrahim Peak. Fifty years ago, the remains of nine Neanderthals were found here. Their fossil bones included an intact skull and several skeletons that are estimated to be about 75,000 years old. They indeed represent one of the most complete fossil remains ever found. Besides a vast collection of bones, paleontologists noticed two other remarkable things in this cave. First, peculiar remains suggestive of elaborate burial ceremonies, such as seeds and pollen of hyacinth and some other flowers around one grave, probably representing a practice similar to our custom of laying flowers over graves of the dead.

Furthermore, the bones of one of the adult males showed signs of injuries to his left eye and right foot. But those injured bones seemed to have healed before he died at an old age of 40 years. It has been postulated that the healed bones of the old man are signs that Neanderthals looked after their injured.

So far, the fossil bones of about 400 Neanderthals have been discovered in over 80 sites in many regions of Europe (except the northeastern part and Scandinavia), as well as western Asia, including the Middle East and also some regions of Central Asia. Neanderthal remains represent the best fossil records among all primitive humans. More than half of these individuals have been found in caves and rock shelters in burial arrangements. Several caves contain the remains of so many individuals that it has been suggested that these caves were indeed their cemeteries and not just typical residential caves.

Originally, Neanderthals were portrayed as ape-like and brutish cavemen. But now it is generally accepted that they did not look much different from modern humans, considering the fact that living humans themselves show noticeable variations in their bone structures from one group to the other. Nevertheless, scientists in this field express quite different views about their relatedness to modern humans, and whether they should have been considered extinct or whether they were actually assimilated with modern humans. Some place them in a separate species, *Homo neanderthalensis*, implying that they were not our direct ancestors, while some others categorize them at a subspecies level, *Homo sapiens neanderthalensis*, with the capacity to interbreed with modern humans.

They were generally shorter than modern humans, on average five feet tall, but had stocky bodies, very robust and muscular, and had big skulls, at least as large as those of modern humans. But their brain cases are easily distinguished by being longer horizontally and shorter vertically with a little bump at the back, called the occipital bun. Their faces, on the other hand, were longer and protruding along the mid-line, especially around mouth, with a large nose cavity, massive brow ridge and a general absence of chin.

It is suggested that Neanderthals were evolved in Europe from populations of archaic humans, such as *Homo erectus*, who migrated to Europe about half a million years ago. Their migration route could have been either from northern Africa through the Strait of Gibraltar in southern Spain or through the Middle East. The European fossil records, roughly ranging in antiquity between 130,000 (the start of last Ice Age in Europe) to 30,000 years old, unmistakably show typical Neanderthal features. Among the last fossil remains of Neanderthals is a mandible found in the majestic Zafarraya Cave in southeastern Spain, estimated to be about 30,000 years old. Then, abruptly, they disappeared from Europe, while at the same time the oldest Cro-Magnon fossil remains of modern humans started to show up nearby in Spain and France.

Due to the Neanderthals' long history in Europe, it is postulated that west Asian Neanderthals originally migrated south from Europe during last Ice Age. Their fossil remains, dating back about 80,000 years, are mainly represented in several sites excavated in present-day Israel. Additionally, their fossil bones were found in Iraq, Syria, and Uzbekistan. The abrupt transition of Neanderthals to modern humans in the European excavated sites argues for the replacement of Neanderthals by modern humans with no hybridization.

However, the situation in western Asia is quite different. The fossil remains unearthed in western Asia represent the best case for the evolution of Neanderthals into modern humans. During the last 75 years, several sites excavated in present-day Israel, particularly two caves in Mount Carmel near Haifa, have provided fossil remains of both Neanderthals and near-modern humans. But the exact dating of the caves has been debated for sometime.

Estimates ranging from 45,000 to 90,000 years make four possible scenarios: first, Neanderthals evolved into near-modern humans; second, Neanderthals co-existed with the near-modern humans originating from Africa; third, Neanderthals in this region replaced the near-modern humans from Africa;

fourth, the near-modern people from Africa replaced Neanderthals.

A few morphological features, such as a sloped forehead and long face typical of Neanderthals are also noticeable in some people of western Asia. Thus, they might point to the continuity of Neanderthals' features in modern humans. Moreover, there is another argument one could make for the possibility that the Neanderthals had some contributions to the genome of modern humans. As mentioned above, an extensive fossil record indicates that Europe, western and Central Asia were the only regions that have yielded the remains of people with Neanderthal characteristics. We also saw in the previous section that the living people of Europe and western Asia are genetically more similar to each other than to the people of other regions (see p.117).

The morphological similarity in the past, and the present genetic similarity between the two groups may not be so just because they are occupying two adjacent regions. It might point out to their genetic relatedness going back hundreds of thousands of years, with the implication that Neanderthals never really died out completely in Europe.

As described in the previous section, recently some limited DNA evidence (see p.162) have been taken to indicate a complete replacement of Neanderthals and other primitive humans by modern humans everywhere. Based on the DNA results, it was calculated that the most recent common ancestor of Neanderthals and modern humans (*Homo sapiens*) had lived about half a million years ago. Consequently this figure has been generally used as an estimate of the origin of the Neanderthals in Europe.

However, other limited data from studies on the human X-chromosome paints a rather different picture about the extinction of all other hominids who were contemporaries of modern humans. Two recently found haplotypes on the X-chromosome seem to have originated in Asia rather than in Africa, implying genetic admixture between modern humans and ancient hominids of East Asia (haplotype is a set of DNA mutations inherited together on a known piece of DNA).

Whether the Neanderthals and other primitive humans were completely extinct or whether we still carry part of them in our genome is currently a question with no definite answer. Meanwhile, in the next chapter, "Who Were *Divs*?" we will see how their legacy is still alive in the collective memory of the people of the Middle East, vividly revealed in some ancient Iranian folklore.

Who Were *Divs*?

A Legendary Mother *Div*

One by one, we crawled in darkness through the muddy waters of a narrow stream carved out in limestone rocks, barely visible with the flashlights attached to our helmets. Then, we entered a small room surrounded by walls made of big boulders with no more passages, thinking it must be the end of our wild cave expedition. But in no time, our group leader looked up at a tiny opening near the ceiling and with a big grin on his face told everybody to move up there. I thought, Jesus Christ, how are we supposed to get up there now? I don't remember exactly how we managed to pass through it, since it was so many years ago. But we did pass through that little hole somehow and many more narrow passages, while getting down on our hands and knees and squeezing ourselves under the massive weight of an old crumbling mountain.

It took more than half a day to finish the expedition course, and it turned out to be one of the most physically challenging adventures of my whole life. It was a cold winter day in February. We entered the cave early in the morning through a vertical opening about 30 feet deep, descending to its floor with a rope attached to a harness around our waists. Then we walked through a vast underground network of caves called Johnson's Crook that lay beneath the Lookout Mountain, part of the Blue Ridge Mountains in the easternmost range of the Appalachians, south of Chattanooga, Tennessee, near the Georgia border. By afternoon when we left the cave, my jeans were so drenched in mud that even after washing them three times, the fabric looked as if it was engraved with red clay, and I had no choice but to throw them away.

It was 23 years ago, and I was a graduate student at the University of Georgia in Athens, a very beautiful highland area of the South. One day, while waiting for the elevator in our science building, I noticed a flyer on students' activity board, announcing a caving trip in nearby Tennessee. So I decided to give it a try. But after that trip, I never could look at a mountain the same way as I used to. My majestic solid mountains suddenly became no more than a chain of hollowed inter-locking chambers.

That was the second time that I felt so deluded. The first was in 1976, in a big lecture hall at Stanford University. I had taken a biochemistry course given by several members of the biochemistry department. That day, Professor David Hogness, a geneticist, was giving us a lecture about the structure of genes. Although he was not a Nobel laureate, like two other professors who were also teaching that course, he was my favorite one.

Nowadays, it is very common to see large stretches of DNA sequences as shown in scientific journals. But 30 years ago, it was a rare sight. I still remember the big screen covered entirely with rows and rows of DNA sequences that he had shown us for the first time in one of his lectures. I also vividly remember when he gently pointed to a nucleotide (the chemical name for the building block of DNA) in the middle of the screen and said, "The actual gene starts from here. All nucleotides before it are just non-genic spacer DNA." That was a radical departure from our previous notion about genes, that is, set next to each other on each chromosome like beads on a string with nothing between them. Now actually, we know that over 97% of our human genome consists of these so-called "junk DNA" with no well-defined function.

That was the day I learned about our hollowed chromosomes, and now I was inside a hollowed mountain. I wondered how many more hollowed giants I would have to encounter in my life. On that trip, I lost my great respect for mountains as solid giant entities, and instead I gained great respect for the human mind, for being so brave, so adventurous by walking through those dark underground passages following the lure of the unknown.

On our return trip back to Athens, Georgia, our tour guides, an American couple in their late thirties, very down to earth themselves, told me about some renowned deep caves in Iran. I thought, so those were probably the caves that I had read about in my childhood, where Ahriman (Satan in Farsi) and his servants, *divs*, dwelled in the underground world in darkness, hiding from Ahura Mazda (Supreme God in the Zoroastrian religion, which was the main religion of Iranians before Islam) and his lieutenants, *fareshteha* (angels).

I remember, as a child, I was so scared after reading those stories that I could not sleep at night. But now, being in the middle of this underground world, supposedly stretching several miles around me, it was not scary at all. It was almost surreal to be there, experiencing the settings of our prehistoric ancestors who once inhabited these mysterious sub terrains.

I always wondered, who were those *divs*, the monster-like creatures with many human characteristics? In the stories I had read in my childhood, they were not just symbolic beings representing the evil forces of our mind. They were real, even more so than *fareshteha*, the angelic beings who were supposedly protecting humans from evil forces. Iranian folklore is full of stories about *div-o-pari* (*div* and *pari*). *Pari* is probably the same as pariah in English, which means outcast. Additionally pariah refers to a member of one of the lowest social castes in India. *Pari* has also been translated as jinn, which is an Arabic word commonly used in Farsi to describe a special creature with the power to disappear out of sight in an instant.

However, *Pari* personifies a completely different type of being among Iranians. Like *fareshteh*, it was supposedly a beautiful and dream-like female creature. *Pari*'s name is also mentioned a lot in conjunction with another beautiful creature called *huri*, as in *hur-o-pari*, indicating legendary beauties. *Fareshteh*, *pari*, and *huri* are still used in Farsi as common names for girls. It is noteworthy to mention that Hurian is also the name of an ancient tribe who dwelled in Syria near the Turkish border.

Div is also a very common word in Farsi, still in use to describe nasty and vicious people. I suppose every human culture has some sort of prehistoric human-like hairy and scary creatures like Iranian *divs*. There are also two common words in Farsi that we use routinely without really knowing that they refer to *divs*. One is *efriteh* meaning an "ugly and extremely selfish and vicious woman," and the other one is *patiareh*, a common mild swear, meaning "whore." Indeed, both these words refer to the names of two female *divs* in the old Iranian folk tales. *Div*, written as diva, *dev*, *dew*, *daiva*, *daeva*, or *deva*, depending on the language it has derived from, as an Indo-European word has a very interesting history.

When Iranians and Indians shared a common heritage going back about 5,000 years, they supposedly worshipped two sets of deities, *ahuras* (lords) and *divs*. However, after they separated into two groups, their deities were also affected. *Divs* became the wrong deities among Iranians, whereas *ahuras* (*as-*

uras in Sanskrit) became the wrong deities among Hindus. Therefore, by the time Zoroastrian religion became widespread among Iranians, *divs* had already been morphed into supernatural nasty demons worthy of damnation, as indicated in their religious texts. There is even a well-known inscription on one of the ancient ruins of Iran in which more than 2,000 years ago, a king of the Achaemenid Empire of Iran, Khashayar (Xerxes in Greek), son of Darius, talked about *divs* as the wrong deities. He declared, "Within these provinces there were places where previously the *divs* had been worshiped. Then by the will of Ahura Mazda, I uprooted that cult of *divs*, and I made a proclamation that the *divs* shall not be worshipped."

However, in Iranian folklore, *divs* are personified not as supernatural beings, but rather like real human-like people with many human characteristics and some supernatural capabilities, who sometimes overcome their opponents by means of magic. They interacted with ordinary people and had personalities, names, families, kings and kingdoms. So it should not come as a surprise that through the ages, *divs* have occupied a very special place in the Iranian literature, as well as folklore.

There is one book in particular that always brings to mind some fascinating characteristics of *divs*. This unique book, which is not old at all by Iranian standards, for several reasons, has managed to occupy a special place in the minds and hearts of all who had a chance to read it. Ironically, few people remember the name of its gifted author(s), or even wonder how this detailed adventurous folk tale woven together so magically and effortlessly did eventually end up as a book. Maybe it is one of those imaginative yet intensely familiar tales that bypasses authorship, as if it belongs to one's own past, going back to our childhood state of mind where everything seemed attainable.

This book was written about one hundred fifty years ago, originally not as a book for sale and public consumption, but as a way to preserve some ancient legendary folk tales. These are the tales that have mesmerized generations of children and grown-ups who have cherished them as if they were the real history of their own folks.

It is one of those rare books that after just a few pages can masterfully transfer the readers to another realm, full of enchanting characters and magical places. The book is called *Amir Arsalan-e Namdar*, "Amir Arsalan the Celebrated." It is the detailed story of an adventurous and romantic prince, often disguised as a commoner, who traveled many real and unreal lands and braved the spells of sorcerers, encountered many *divs*, dragons and politicians to rescue his beloved and headstrong princess.

Although the storyline sounds very typical, the details are so original and captivating that I read somewhere that a noted Japanese novelist has recently reworked this story into Japanese. He started his epic work called *Heroic legend of Arslan* in 1986 and as of 2004, he has extended this story into eleven volumes, though not completely finished yet. The response to his efforts had been so great that in the '90s, his story was first turned into comic cartoon books, and later into a series of Japanese animation works.

One of the most fervent admirers of this story was a 19th century Iranian king named Nasereddin Shah, the most renowned king of Iran's Qajar dynasty. In those times, folk tales were not written down as books and were usually narrated in coffee shops by enthusiastic and professional storytellers as a common means to entertain ordinary people.

Nevertheless, Amir Arsalan was more than a collection of ancient Iranian folk tales. The storyteller of the king's court, the genius-minded Naghibolmamalek, crafted the story especially for him. He knew the king very well, who around one hundred fifty years ago showed much interest in everything European. Consequently, the storyteller created a princess who was the daughter of a foreign king, and many of the adventures of Amir Arsalan happened in foreign lands as well; while at the same time, the name of the princess remained Iranian and was called Farokh Lagha.

Apparently, one of the duties of the storyteller of the king's court was to narrate the Amir Arsalan story at least once a year for the enthusiastic king. The court arrangement for nightly storytelling was very interesting by itself. As it goes, on the second floor of an inner palace building, there was the king's private quarter opening into three adjacent rooms. One room was where the storyteller and his musicians would sit and through an open door would narrate his stories for the king. The other two rooms were assigned to the king's special guards who would assemble at night to protect him from intruders. Not a bad arrangement.

It is known that one of the king's daughters was very interested in books and had her own library in the palace. Her official title was Fakhroddoleh (pride of the court) and she was really interested in turning Amir Arsalan's story into writing, probably to add it to her vast literary collection. So with the king's permission, she made an arrangement to sit every night with the king's guards in one the adjacent rooms, with the door half open, her small wooden desk in front of her, writing down the storyteller's adventures of Amir Arsalan.

The royal father and daughter were supposedly so fond of this story that on occasions when Fakhroddoleh had to stay in her own house for the night, the king would ask the storyteller to tell another story, so his daughter would not fall behind in writing the Amir Arsalan story.

It is very hard to say how many lines of the final manuscript, which is more than 600 pages long, were originally told by the storyteller and how many were inserted later by the king's daughter. I remember when for the first time I read this story, thinking that it was written by a guy, I was really puzzled, when reading many passages, by how well a guy, in those days, could perceive the delicate feelings of women in general, and the willfulness of princesses in particular.

One of the key events of the Amir Arsalan story is his encounter with *Div-e Fooladzereh* (the *div* with an armor made of steel), a monstrous and giant creature living in a deserted castle. But the poor *div*, as described routinely in other folk tales as well, was desperately in love with a beautiful maiden of human race. Although brutish by nature, being in love, he was very gentle and affectionate towards her, singing for her and praising her beauty, and tenderly asking for her hand in marriage, while the woman relentlessly shed tears in vain.

A special feature of *Div-e Fooladzereh*, as typical of other *divs* in other folk tales, is the way he commuted, in a distinct aerial wandering described as "throwing wind/ air in furnace and going up in the air," which actually nobody knows how it worked. It might have involved a special way of breathing and enlarging the lungs.

Another famous character in this story is the mother of *Div-e Fooladzereh*, commonly known as *Maadar-e Fooladzereh*. She had spellbound her son by making him a talisman to protect him from any harm, except one, a special sword ornamented with many pieces of emerald. In many ancient myths, among different jewels, the emerald was the one with magical effect in offsetting the effects of evil-minded creatures. Thanks to her vividly described personality in the Amir Arsalan story, *Maadar-e Fooladzereh* has become like an adjective in Farsi language. Even those who have not read the book or not heard anything about her personality as a *div* know exactly what the word means: a big domineering woman, a female bully.

The *Div's* features are pictured in detail in many books. They are typically depicted larger than humans, but not awfully larger, and usually have black or red skin which is sometimes spotted. They usually have two horns on top of their heads, tufted tails, like those of lions, and claw-like toenails. They have beast-like faces with sharp fangs, black lips; and the white portions of their eyes look more like a darkish blue. Nevertheless, instead of looking ferocious, they come across as funny, with those short skirts that they are usually wearing, and with their somewhat sweet faces.

Thirty-Thousand-Year-Old Stories

Since Amir Arsalan is mostly an imaginary story, it could be argued that its characters, including the *divs,* were made-up figures. However, it should be mentioned that *divs* are also represented prominently in scholarly works, such as *Shahnameh*, the record of kings, an epic poetry book, written about 1,000 years ago, by the Iranian poet, Ferdowsi. This book was compiled to bring to life not actually the historical but the mythical kings of Iran. Nevertheless, it was based on extensive documented research on ancient Iranian history, and some of its characters are similar to those mentioned in Avesta, the holy book of Zoroastrians, and also in Vedas, the ancient religious texts of Hindus.

According to *Shahnameh,* the first *div* was the son of Ahriman (Satan). Early on in this book, there are constant wars between *divs* and the armies of various Iranian kings, until one of them, called Tahmoreth, finally conquered the army of *div*s. After getting captured, they asked him not to kill them, and in return they taught him how to read and write. During the reign of his son Jamshid, a legendary figure in Iranian mythology, *divs* and *pari*s were brought under his command again and built houses and other buildings for his cities. Then, as time went by, there is less and less mention of the *divs*, and only occasionally there has been a new encounter with them, as if they had been either assimilated in Iranian society, or vanquished altogether.

The most renowned *div* of *Shahnameh* is Div-e Sefid-e Mazandaran, whom Rostam, the hero of this book, had a significant encounter with. Mazandaran is the name of a province in the north of Iran. *Sefid* means "white," as he was depicted as a very strong *div* with black skin and white hair. He is mentioned in a story about Kaykavoos, a mythical Iranian king, who with a big army went to war with the king of Mazandaran. So the latter king sent a messenger to Div-e Sefid who lived in a far away place, inside a dark and deep cave, asking for help.

The *div* agreed to encounter the Kaykavoos army, and during nighttime arranged for a hail of stones to fall over his army and thus defeated them in the first round of fighting. Then, according to the book, after making them prisoners, Div-e Sefid put 12,000 *divs* in charge of the prisoners and assigned four *divs* as their generals. Afterwards, he went back to his secluded cave. From the earlier passages of the book, we had learned that Rostam, who was consequently called upon to rescue the Kaykavoos army, was looking specifically for Div-e Sefid, as well as the other four important *divs*, each with his own specific name.

Another well-known *div* whom Rostam had heard about and was aware of his menacing actions was Akvan-e Div, who resided near a spring. The *div* considered the vast plains around the spring as his own territory and whenever the nearby shepherds with their herds were approaching for grazing, he would get into action. To evade the shepherds, he would dress in zebra skin, as if he was a zebra and would relentlessly attack horses and sheep alike.

It seems clear from the details and narration of these stories that these *divs* were not some sort of evil spirits, but rather a different kind of people with their own way of life. Interestingly, in *Shahnameh*, *divs* were generally spoken of as groups or populations of people with specific characteristics, whereas in folk tales they appear more as novelty and individual cases.

One reason could be due to different time periods in which *divs* appeared in various Iranian myths. Ferdowsi has boasted in his poetry that he is narrating 30,000-year-old stories. Thus, if *divs* were indeed some indigenous people originally inhabiting the Iranian plateau, they would have lived in large groups of people some 30,000 years ago. It should be mentioned that there is no indication why Ferdowsi chose to suggest a date as far back as 30,000 years. It is quite rare in Iranian literature to talk about such a distant prehistoric time with so much assurance.

Folk tales, on the other hand, represent a more recent time than 30,000 years ago. Therefore, probably by then, *divs* were representing some vanishing people, appearing sporadically among other people and capturing their attention. In another scenario, *divs* might have genetically admixed and culturally assimilated with modern humans a long time ago. So, the *divs* that appeared in folk tales might represent small packets of the original *divs* who never got mixed with other people and lived in remote areas by themselves, coming into contact with regular people only once in a while.

Typically in folk tales, *divs* were portrayed as brutal but simple-minded people. One of their unusual characteristics was that they usually would act contrary to what they were asked to do. So if the hero of the story wanted to be thrown into the sea, rather than into the mountain, he would ask the *div* to do the opposite action, that is, throw him over the rocks of mountains. It is said that they had a tendency to feel sleepy in sunlight and usually slept during the day, typically in dry wells or inside deep caves. Sometimes, they would sleep for several days in a row. This habit usually would be used by their opponents to approach them while asleep, and attacking them after waking them up. Their presence would be typically noticed by a fowl smell in the air or a sudden increase in temperature.

Those *divs* who were overcome by a king or hero would often serve him as his slave. In one story, after the defeat of a *div,* his ears were pierced by the victor and two horseshoes were inserted as the mark of the *div*'s servitude. A common theme in many stories is the *div*'s offer to wrestle a hero and saying, "If you win, I will be your slave," and thereafter would serve him very faithfully. Usually, they would leave a bit of their hair with the hero to put it in fire when their help was needed.

As mentioned earlier, they were very fond of human females. Typically in *div* folk tales, a *div* would steal a woman, take her to his home, which usually was at the bottom of a dry well, place his head on her lap and sleep there for several days. Although the unlucky bride would usually try to escape this most unfortunate fate, in some stories we were told that the bride was so taken with the *div*'s affectionate manners that she would defend the *div* when confronted with her own family.

I remember when we were children, during lazy summer afternoons; we were usually kept quiet by stories of *div-o-pari*. Usually for us the most exciting part was when the storyteller would get to the point when it was the time to find the life bottle of a *div*, a rather small glass bottle that contained his life. Usually that was the only way to kill a *div*, to find and break his life bottle. In some stories *divs* had several heads, and sometimes they had only heads with no bodies and moved around by rolling their heads. That was because their opponents beheaded them, but still their life bottles were not broken.

In some other stories, instead of the life bottle, a *div* had an external soul, and likewise, destroying it was the only way to kill him. The external soul was usually hidden by the *div* in a box, in a live fish, or in the body of some other animals, such as a mountain goat. Thus, as soon as the mountain goat would be killed, the *div*'s external soul would be destroyed, turning him/her into smoke and making him/her disappear into the air. The concept of external soul associated with *div*s in these stories is extremely intriguing, and as will be discussed later in the chapter, "Who Might We Be?" might fill a big gap in the later stages of human evolution.

All these folk tales could at best be considered some highly exaggerated accounts of the lives of a vastly different group of people who probably, in the very distant past, shared the same surroundings with our ancestors. Or possibly they represent a collection of strange behaviors originated by many different types of people who had lived a long time ago, and then were collectively attributed to only one group called *divs*. But the most striking reminder of their legacy, as simple-minded people, is distilled in one simple word in Farsi, and the only word that Iranians routinely use to describe people who are fool, crazy, or insane.

The word is *divaneh* and it exactly means "like *div*." It has three meanings. It means having a mental circuitry that erupts into anger very fast. It also means to be crazy about someone or something. And on other occasions it could mean someone with lower mental capacity. So it refers to a person who could be angry and violent like a *div*, or be in love like a *div*, or be a fool like a *div*. All these three cases are the personifications of *divs* as reflected in the above-mentioned folk tales. Some have also suggested that *divaneh* could mean, "being possessed by a demon," but this is not really the literal translation of the word *divaneh*. According to Farsi grammar, the addition of *aneh* to many words would affix "like" to their meanings.

So who were *divs*? If Ferdowsi were correct and *divs* lived some 30,000 years ago, who could *divs* be? Ironically, the presumed basic physical and mental characteristics of *divs* are surprisingly similar to those routinely described for the Neanderthal people. The estimated age of the unearthed bones of the last Neanderthals, who once roamed Europe, the Middle East and Central Asia also goes back 30,000 years. What a coincidence!

Who Were Adam and Eve?

Ancient Tales of Mandaeans

I used to go up there very early in the morning, usually on Sundays, about seven o'clock, before buses packed with tourists from all over the world would arrive. It was some 23 years ago. At the time, I was doing my first postdoc in the Genetics Department of the University of Hawaii at Manoa, in Honolulu, Hawaii, a very big and beautiful campus. To me the Poli Lookout, in the outskirts of Honolulu, was one of the most serene places on earth. It was a large man-made platform perched on a mountainous area, always windy and usually misty. On one side, it was surrounded by trees, and on the other side, there was the panoramic view of a vast landscape, with scenery absolutely out of this world.

As far as eyes could see, there were virgin lands stretching everywhere and merging with the Pacific Ocean in the distant horizon. It seemed that the opulent scenery was not showcasing any land or sea; rather its magical shades of blue, green and purple were intended to reveal nature's timeless inner peace.

That scenery would always remind me of a much earlier time, before earth became so crowded with so many people. One could imagine, out there, early humans in small groups wandering around those green acres, pursuing a simple life.

Now, as I am living in San Francisco, nothing seems so innocent, so primitive anymore. Here, it rather seems like a microcosm of the convergence of all humanity in one place. Usually, when I walk out of the BART Metro train station into crowded downtown San Francisco, I tend to forget that I am an Iranian supposed to be an American after living in this country for over 30 years. In downtown San Francisco, one naturally feels that one belongs to the human race, an international citizen. There is always something interesting going on in downtown San Francisco: a typical parade or, more often, a weird one, a political demonstration, or some street performers entertaining the crowd. And of course, whether shine or rain, there are always those tourists, some from other countries, some from other states, waiting in long lines for cable cars to take them down the hills to the Fisherman's Wharf.

But the international feeling that local people like me experience comes from being in the vicinity of other locals who are from all over the world, especially Central America and Asian countries. To me, as a geneticist, nothing is more fascinating than looking closely at the moving sea of people crossing the busy streets. I get the chance to look at their noses, lips, eyelids, eyebrows, foreheads, chins, cheekbones, even their ear lobes, trying to figure out what would be the features of a human prototype face. However, that is not an easy task for me because my natural tendency is to see a crowd in each face rather than to extract a typical face from a crowd.

It is very riveting to see many people in one face, as if the whole world is made of close relatives. I remember, when we were children watching TV, rather than following the storyline of the movies, my sister and I had much more fun concentrating on the actors' faces, trying to figure out who they reminded us of.

Needless to say, not everybody sees the world that way. Some 20 years ago when I was living in LA, I used to regularly visit the Church of Latter Day Saints (the Mormon Church) in the Westwood area to look at the books in its extensive genealogy library for my birth date study (see p. 92). That was before the Internet age, and I had to copy each book page after page for an extended time. While standing there copying the books, it was hard not to notice the puzzling looks on the faces of middle-aged women around me, all of them of European descent. It seemed to me as if they were wondering–what is she doing here, copying book after book? Her ancestors are definitely not in these books. Why is she wasting her time like this? And I kept smiling at them thinking, they are right; my ancestors are not there. But I was not looking for my ancestors; I was looking for everybody's ancestors.

Who are our ancestors? Obviously we evolved from higher primates, and probably we and our distant cousins, the great apes, had common ancestors some 5-7 million years ago, but what about Adam and Eve? The Adam and Eve best known to the world through the Old Testament. Were they real people?

On Sunday mornings on the bank of the Karun River in the city of Ahvaz in southern Iran, a small group of people gather for an elaborate and ancient baptizing ceremony. All dressed in light white robes and turbans, they sit in an orderly line or two perpendicularly to the riverbank. Before the start of the ceremony, an old man with a long white beard, the baptist, sits by the river and quietly makes a simple traditional banner called *drapsha*. First, he lays two sticks, made from olive tree branches, on his lap; he puts the smaller stick over the longer one in a cross-like shape, then he fastens them together with a branch full of small green leaves. He sticks the cross in the sands of the riverbank and wraps it with a long light brown cloth. Next, he burns incense in a small and ancient-looking clay burner.

Then, standing next to the small cross, he and two other bearded guys start reciting prayers from their ancient holy book to purify the air. They read with the cross's long cloth covering their hands and supporting their books, laid wide open in front of them. The long cloth is stretching from them to the delicate cross, unifying all of them while facing their beloved flowing water glittering under the bright sun.

In their appearance and their simple yet profound manners, they are reminiscent of John the Baptist whom they regard as their last prophet, and whom they know as Yahya bin Zacharia, Yahya the son of Zacharia. But make no mistake, they are neither Christians nor Jews, and ironically in the currently Islamic Iran, a country with a Zoroastrian past, they are neither Muslims nor Zoroastrians.

They belong to an ancient and forgotten sect called the Mandaean Sabians. Mandaean is their tribal name and *sabians* means "those who immerse themselves in water." The word *manda* in Mandaean is believed to mean "knowledge." Mandaeans number about 5,000 to 25,000 in Iran. Some of them speak their own language, Mandaic, one of 75 languages currently spoken in Iran. Mandaic is a dialect of the eastern Aramaic language, which is a member of the Semitic family of languages. Aramaic was widely spoken in the Middle East during the Achaemenid Empire of Iran about 2,500 years ago. Mandaic sounds like a mixture of Arabic and Farsi.

According to their teachings collected in numerous books and scrolls, and written with their own unique alphabet, during their long history, they had lived in Egypt, Jerusalem and in northern Iraq near the Euphrates River close to Turkey. In addition to Iran, an estimated 30,000 to 70,000 of them currently live in Basrah in southern Iraq near the Tigris River. However, they do not look either like Arabs or Jewish people. From the pictures I have seen, with their fair skin and other facial characteristics, they look more like the people living in western Iran and northeastern Iraq, which cover the areas they supposedly lived previously. Despite their simple way of life, these people are not nomadic, and they are indeed highly respected goldsmiths and jewelry makers, following their ancient ancestral trade. Although in Tehran few people have heard anything about Mandaeans, they are known among jewelry-conscious women as the goldsmiths of Ahvaz, who make antique-style jewelry.

Recently, this sect has received much attention from the Christian scholars who are intrigued to know more about John the Baptist and also about Gnostic traditions. Likewise, Muslim scholars have shown interest in the identity of Mandaeans for a long time. In the Quran, three groups of people have been identified as having religious books: Christians, Jews and Sabaiyans. It seems that Sabaiyans were well known at the time of the Prophet Mohammad, but not anymore. Nobody knows for sure who they are now. Currently, there are three groups who most likely represent them, one of them being the Mandaean Sabians. And that is how I learned about them three years ago, after reading the Quran for the first time.

Although Mandaeans are followers of John the Baptist, they believe that Adam, or as they call him with the highest respect *Hazrat-e* Adam, was their first prophet and teacher. Their holy book, *Ginza Rabba* (the Great Treasure), contains many details about Adam, whom they believe was the first man. They believe that Mandaeism was the original religion of Adam, and that their holy book was revealed to him, from *Haii* (Life, the Creator) through *Hibil Ziva*, one of their five major *Uthras* or angels.

They also believe that they are direct descendants of Hibel, Shitel and Anosh (Anus) who were the three sons of Adam and Hawwa (Eve). Hibel seems to be similar to Abel of the Old

Testament. They consider Shitel, who is the same as the Biblical Seth, the third son of Adam and Eve, as their second prophet. Anosh, still a familiar name in western Asia, including Iran, as in Anoshiravan, meaning "Anosh-like soul," sounds like Enosh (also called Enos, meaning "mortal man") of the Old Testament; although in the Jewish text, Enosh is the son of Seth and not his brother. Their third prophet is Shem (Saam), son of Noah. And their fourth and last prophet, as mentioned before, is Yahya bin Zacharia, commonly known as John the Baptist.

According to them the three sons of Adam lived among 60 brilliant angels who had come to earth to be their mates and bear their offspring. The texts of Mandaeans are full of details about the brilliant angels who came from the "world of light" and assisted in specific tasks related to Adam and his offspring. Their most important and powerful angel is *Manda-ed-Haii*, meaning the "knowledge of life," to whom they always address their prayers. This angel is supposedly endowed with the power of the Creator. It is in constant fight with *Ruha*, a feminine owner of the world of darkness (see also p. 271).

According to Mandaean oral traditions and folklore, at first *Ruha* and her son *Ur* tried to make Adam. But the Adam they made had an ape-like face and moved like apes as well. So they went to the house of *Haii*, the Creator, and other angels, and told them about their failure. They said they would send *Hibil Ziva*, who came down to earth with the soul on his hands. When the soul saw Adam, she was horrified and only later when she was promised a world on earth, just like the "world of light" with flowers, running water, and pure air, she consented to enter the body of Adam. Right after, Adam stood up and started to talk.

Mandaeans believe that the human is made of three parts: body, spirit and soul. Spirit is a sign of the material world and of evil, whereas the soul is a sign of the "world of light." In their holy book, *Ginza Rabba*, Adam recalls how he acquired his soul. First, he talks about how an *Uthra* (angel) accompanied him from the "house of life." Then *Uthras* washed him in the Jordan River and baptized him, then strengthened him with their pure words. Later, on another occasion, he describes how one of the great *Uthras* took a circlet of radiance, light and glory, and set it on his head. As mandated by *Manda-ed-Haii* in the ancient time, every man of this sect must wear a light white turban symbolizing the great mystery of radiance. In addition, as part of the baptizing ceremony *eklil*, a very small crown or wreath of jasmine or myrtle is placed under the white turban of the baptized, above the forehead, as a symbol of intellect.

The Mandaean calendar is based on Adam's creation, and according to them we are now in the year 445,369 of Adam's birth. Coincidently this date is very close to the date estimated by mitochondrial DNA studies, about half a million years, as the date the most recent common ancestor of Neanderthals and modern humans lived in southern Europe. Or if one considers 445,369 not as in years but as in months, it would be about 37,000 years ago, which coincides with the time the bones of modern humans started to show up in many European caves. In either case, this numerical value, whether real or imaginary, could have been known only through *elm-e ghaib* (Gnostic knowledge, see p. 68), which is the basic tenet of all Middle Eastern beliefs.

A few years ago, I was sitting in my living room, deep in thought about something, while at the same time staring at the floral patterns of a Persian rug beneath my feet. Then, I noticed how similar the general design of these kinds of rugs is to the structure of a typical biological cell. The particular design of Persian rugs goes back a few thousand years. In its original design, a rug consists of three sections: a border at the periphery, a medallion in the center, and a space between these two, usually filled up with small floral motifs. This is very much the same as the general architecture of a cell. It consists of a plasma membrane, a large and round enclosure in the middle called a nucleus that houses chromosomes (genetic materials), and cytoplasm that fills up the space between the membrane and the nucleus. Cytoplasm is filled with small organelles.

So far, the similarity in the general design of these two items could be considered as being just coincidental. However, the similarity between the details of the plasma membrane and the rug's border is really remarkable. The rug's border consists of three parts itself. There is a narrow outer border with a simple floral pattern. Then, there is a narrow inner border exactly with the same pattern. The space between these two borders forms a middle border, which is two to three times larger than the other two borders, and it is filled with large and elaborate floral motifs.

Looked at cross section, the cell's plasma membrane also consists of three layers. A thin outer layer and a thin inner layer looking exactly the same, and a much larger layer between them filled up with big chunks of materials. Chemically speaking, the two thin layers and the area between them are called the lipid bilayer, which is composed of many units of a single lipid structure with a non-polar head and polar tail. The polar tails stick together and form the middle layer, while their non-polar heads repel each other and make up the two thin layers at the top and bottom. The big chunk of materials in the middle are integral proteins, many of them acting as cell surface receptors (specific chemical structures that make contact with the cell's outside domain).

Obviously, the ancient designers of Persian rugs did not know anything about cell biology. That is what we call having *elm-e ghaib*, being aware of nature and its particular designs and patterns without any direct effort in learning its secrets. Western philosophy has consistently rejected the Eastern concept of *elm-e ghaib*. It seems such nonsense to the majority of Westerners. For example, regarding this particular pattern of the Persian rug, one could reasonably argue that most probably the design was gradually evolved through time and was not inspired by one individual designer. Thus, its similarity with the cell structure is just a coincidence. There are, however, other cases that would easily show the amazing capacity of the human mind to master a great amount of knowledge and awareness with no apparent effort.

One particular case is a mental disorder called autism. About 10% of people with this disorder have amazing memory for a particular topic of their interest. As was depicted in the movie *Rain Man*, the main character, who was mildly retarded and had autism, showed an astonishing memory for specific events. He could recall exactly what the weather was like on any given day during his lifetime, or he would know the exact quantity of a big cluster of items without ever counting them. One of the main differences between autistic people and regular people is thought to be that the neuronal circuitry of these individuals is somehow wired differently. So it is quite possible that each of us is endowed with a great amount of innate knowledge that is not routinely accessible to us. Furthermore, it is suggested that prolonged solitude and meditation has the capacity to gradually change our neuronal wiring, enabling us to deeply delve into our state of total awareness.

Adam's First Wife

Besides Mandaeans, Jews, Christians, Muslims and Bahais all consider Adam as their first prophet. In the Old Testament, even a genealogy of Adam is given, which ends with Noah, nine generations after him. Interestingly, since according to Biblical stories, most of the early descendents of Adam, like

Adam himself, lived to be over 900 years old, Adam lived long enough to see Noah's father. Thus, witnessing eight generations of his descendents. Furthermore, as in many other old stories from the Middle East, they had their offspring many, many years after they were born. Adam, for example, supposedly had his third son, Seth, when he was 130 years old.

The story of Adam and Eve is described more or less in the same way in both the Old Testament and the Quran: they ate the forbidden fruit and were driven out of the Garden of Eden. On the other hand, they come to life much more vividly in two other books called the first and second book of Adam and Eve. These books are part of a collection of historical Biblical works that are considered to be fiction, or rather based on folklore. There is no information about who wrote these books or when they were written. It is postulated that an Egyptian wrote these two books about 2,000 years ago. The original copy was in Arabic and not in Hebrew. Apparently, these books were very popular during the Middle Ages.

In the first book, Adam is portrayed as a guy having trouble adjusting to the harsh life outside the Garden of Eden. Every day, he found a new reason to go to the gate of Eden telling God what a terrible life he was encountering outside the garden. One day the grass was too bitter to eat, and another day his cave was too dark to be able to see Eve, and so on. At the same time, Eve behaved very patiently and all the time tried to calm him down, persuading him to adjust to their new lives.

The most revealing part of these books is that they make it clear that Adam and Eve were living among other people/creatures at the time. Indeed, in some oral traditions of Islam, it was suggested that Adam and Eve were not the first man and woman, rather they were the first two intelligent people. According to these non-Quranic Islamic narratives, Adam was a very, very tall man. His height was supposedly several times the height of a regular man.

In Hebrew, Adam supposedly means "soil"/ "light brown," and it is well known that according to some religious texts, God created Adam from dust/soil. In *Tarikh al-Tabari* (Tabari History), a 39-volume historical resource written by an Iranian historian of the 9th century, a specific time has also been given for this creation process. According to this narrative, after God shaped Adam from soil and blew his breath into him, it took 40 days for Adam to come to life. Now this sounds more like a biological experiment requiring a period of time for completion. Nevertheless, the choice of 40 days in this story could have come from another mystifying, yet quite familiar source, that is, the 40 days of mourning customarily observed in the Middle East after a person dies.

For many, the most bizarre aspect of the Biblical Adam and Eve is that God shaped Eve from one of the ribs of Adam. For a long time, this scenario was indeed such a weird notion. But now, with the advent of genetic cloning, it does not sound as unrealistic as it once did. Basically, it means that Eve was a clone of Adam, with a minor technical problem. With our current cloning procedures, in principle, one's clone will have the same gender as the person it was cloned from, and would not be of the opposite sex.

According to some rabbinic literature, it seems that the rationale for God's action might have been to make a more compatible mate for Adam. In the book of Genesis, it is described that God created man in his own image. On the very same line as this description, it is also said, "Male and female created he them" (literal translation). Since this line is from a section prior to the section describing the creation of Eve, some Jewish scholars have taken it as an indication that Eve was not Adam's first wife.

This interpretation is further supported by a fable in a medieval non-canonical Jewish book called *Alphabet of Ben Sira*. According to this folk tale, at the same time that God created Adam from soil, it also created a woman from soil and called her Lilith. Later, after arguing with Adam, she left the garden, mated with several demons and had many children. It should be mentioned that the name Lilith, in the form of Lili and Layla, is still a very common name in the Middle East, where many common Biblical names are not known at all.

This story again verifies the point that at that time, there were other human-like creatures living outside the garden. In recent Western literature and movies, demons are usually portrayed as supernatural creatures perceived only by certain people. Whereas, as discussed in the previous chapter, in Middle Eastern folklore, they were viewed as human-like creatures with some supernatural abilities who lived among normal people and were commonly called *divs*.

Who Were Guardian Angels?

Land of Prophets

It is undoubtedly the most unlikely place in San Francisco. One would expect seeing such a building along a roadside in a faraway oasis, not here in the heart of a fancy city. San Francisco's most magical and endearing place is a massive looking, yet relatively small adobe structure with very thick brick walls, brushed uniformly in white color, sitting under a heavy dark brown tiled roof. Situated in the middle of Dolores Street, one of the most elegant streets in the city, it looks definitely out of place. Dolores Street, an old, very long tree-lined boulevard, goes up and down, faithfully following the hilly curvature of San Francisco. It is dotted with upscale, old Victorian houses overlooking a lush median strip adorned by a row of towering regal palm trees hovering over passing cars.

Historically, the building is the oldest structure in San Francisco, although it seems ageless in comparison with the other old buildings in its vicinity. The street and the building are both majestic and beautiful, but somehow the stately street and the stunning adobe building don't seem to belong to each other.

They are within each other, but distinctly apart—like two separate states of the mind. As soon as I entered the building for the first time many years ago, probably in 1989, I felt as if I had traveled the time tunnel and landed somewhere out of space, a million light years away from the artistically materialistic world of elegant San Francisco. The air filling the inside quarters was eternally charged with the memories of lost anticipations, like those of an orphan desperately searching for his mother in a dream, or a soul flying solemnly seeking its lost nest. One feels so much at home there.

The sanctuary, commonly known as the Mission Dolores, is called Mission San Francisco de Asis, one of 21 missions built during the 18th and 19th centuries by Spanish padres of the Franciscan Order along California's El Camino Real Road. Like the other missions, Mission Dolores has a small cemetery inside the courtyard of its main building. I think it is the only known place in San Francisco where dead people are still allowed to reside in a city too expensive to honor its dead among its living.

While walking over a narrow grassy path between the greenish tombstones of old graves in the cemetery, the voice of my little niece, then five years old, while talking to my father, echoed in my head, "Grandpa, who is this *khoda* that the older kids talk about all the time?" Indeed, in our daily conversations, we Iranians frequently use the word *khoda*, meaning God in Farsi. Actually, I never knew we used it so often until a few years ago, when my Japanese friend asked me, "Why do you Iranians use the word *God* so much?" That was the first time I realized that we must use it a lot, but still I had no answer to her question.

Looking back, I remember my mother was always swearing to "prophet," but I am not quite sure which prophet she actually meant because sometimes she also swore to "124,000 prophets." That is a lot of prophets. No wonder Iran is called the "land of prophets." Then the stern voice of my grandmother from my father's side started resonating in my head, warning her grown-up sons frequently to "Be afraid of God." In those days I thought that was her personal interpretation of God and had nothing to do with the God in our prayers who we addressed as the "merciful." However, after reading the Quran three years ago, I could see for myself why she was so adamant about God being stern and unyielding. One day, I was talking about this issue, on the phone, with my younger brother who lives in LA and has read the Quran, as well as the Old and New Testaments. He amusingly responded, "If you think God is stern in the Quran, just wait until you read about the God in the Old Testament."

Even more intriguing, in the Old Testament, God was typically behaving as if it was a human. According to the book of Enoch (Idris in the Islamic version), God even walked with Enoch, one of the early descendents of Adam. I always thought that God was a spiritual being, something unattainable. But as I learned later, God, as described in various Abrahamic traditions of the Middle East, seems to be rather tangible and interactive. A personal God rather than a universal God. That realization is still very shocking to me.

Many years ago, I got the same unexpected mental jolt about another seemingly supernatural subject—the heaven. It was 1984 and I was a postdoc in the Genetics Department, University of Hawaii at Manoa, in Hawaii. One day, I was driving my mother, who was visiting me from Iran, around the Oahu Island. Our surroundings were so peaceful and heavenly that they inspired her to ask me a very unusual question. She said abruptly, "Do you think the spirit of grandma is here?" I was so shocked by her question that I just responded, "Why do you think her spirit should be here?" She said, "Well, it is said that Hawaii is the paradise, so her spirit must be here." I thought, what a reasonable conclusion, while at the same time I wondered how my mother and I were different from each other. It never ever occurred to me that heaven could be a physical place on earth. I always assumed that heaven was a state of mind or in another dimension of reality. I felt rather embarrassed that my imagination could not come up with something so simple and realistic. I suppose it had to do with my scientific mindset that every concept has to be dealt with as a mystery waiting to be deciphered through our scientific reasoning. My mother and I both could be very wrong about the nature of paradise, but at least her idea was more natural and straightforward. If heaven was or is a place on earth, as could be assumed from the concept of the Garden of Eden, as mentioned in the Old Testament, then there is no reason to think that God has to be only a supernatural being either.

The Elusive *El*

In the chapter "Who Are We?" I briefly talked about how our languages have partially guided us in finding out more about who we are as far as our ethnicity is concerned. Now let's find out how our languages had defined, in the ancient times, who the God and angels were? In Hebrew, as well as in Aramaic, another Semitic language very similar to Syriac (a language of the Syrian people) that was widely spoken in the ancient Middle East some 2,000 years ago, God is called *Elohim*, the plural form of *Eloah*. These two words apparently relate to the Semitic word *El*, which according to some ancient tablets unearthed in northern Syria was the Supreme God of the Levantine region of the Middle East. As recorded in these tablets, more than 4,000 years ago, *El* was at the top of a list of gods, as the father of mankind and all creatures.

Furthermore, the Old and New Testaments are full of names ending with *el*. Just a few examples are: Daniel, meaning "my judge is God," Muriel, "fragrance of God," Gabriel, "man of God," and of course, angel, the most widely used word ending with el, meaning "messenger of God." *El* also relates to another Hebrew word, *Elyon*, which means "higher," "upper." Thus, *El Elyon* is translated as "God the Most High." In Farsi a similar sounding word to *El*, *il*, means "tribe," as in one's origin. There is still a province in western Iran called Elam or Ilam, apparently related to the ancient Elamite dynasty in west and southwest Iran, historically representing the earliest dynasty in ancient Iran, dating back over 5,000 years, and also one of the oldest recorded civilizations in the world.

In Arabic, a common word starting with *el* is *elham*, meaning "revelation," which is also a common name in Farsi for girls. As mentioned before, in Farsi we call God *khoda*. Since Farsi and English are both members of the Indo-European family of languages, God and *khoda* are linguistically related to each other, as their rather similar pronunciations imply. Quite interestingly, a similar sounding word *khod* means "self" in Farsi.

I should mention that when addressing the Almighty *khoda*, we usually call it *aiy khoda-ye-asemanha*. It literally means "you God of the skies," which is similar to the meaning of *El Elyon*, as an "upper God." So, the former phrase was originally either referring to God as a tangible being residing in the sky, or was rather used to invoke *Asman*, a prehistoric Iranian deity, which now literally means the "sky." Indeed, this last example represents one of those seemingly simple phrases that one uses constantly, assuming complete understanding of its full meaning until further reassessment makes it rather confusing and opaque. The same problem is also encountered when trying to understand very old texts. A text may say something that we could recognize word by word, but nevertheless; we read it with a mindset that is totally different than that intended to be its original meaning.

Going back to *El*, in ancient Greek it was identified with *Cronus*, which in turn is identified with the Roman deity *Saturn*, whose name is still in everyday use as in Saturday. In Farsi, we call the planet Saturn *kayvan*. The *kay* part is a common prefix in many ancient names, such as *kayhaan* meaning "universe" and Kayanian (Kianian), an ancient Iranian mythical dynasty brought to life in Ferdowsi's epic poetry book called *Shahnameh*. The name of all kings from this dynasty started with *kay,* as in Kaykhosro and Kaykavoos. These names, with or without *kay,* are still common first names among Iranian boys. So what does *kay* mean? In Farsi, we have another name for boys called Kayumarth (Keyumarth), made up of *kayu* and *marth*, which linguistically has been defined as "mortal life," *marth* meaning "mortal" and *kayu* "life." Notice the similarity between the pronunciation of *kayu/kay* and that of the Greek goddess *Gaia*, meaning the "living earth."

On the other hand, *van* in *kayvan*, the planet Saturn, is an ancient Iranian word meaning "soul," as in *urvan* meaning "primal soul." When putting all these together, the meaning of *kayvan* becomes very intriguing, somehow relating to life and soul. Of course, the planet Saturn, as the upper or seventh heaven of the ancient seven heavenly bodies, had a very special place among ancient people. But it is still intriguing to know whether there was really a connection between Saturn or other planets, and any life form and soul.

In Ferdowsi's *Shahnameh,* although the stories and characters seem to be mythical and at times magical, they all appear real. After all, according to him and others, they were based on well-researched Iranian folklore, representing the people and events of a very distant past. He frequently talks about *divs* as human-like beings interacting with humans. Nobody thinks *divs* are real now, but the way he characterized them was such that one assumes they must have been real beings living in the distant past. Then, out of the blue, in a few places, without any elaboration, Ferdowsi talks about *aflakian* usually as in *khakian* and *aflakian*, as if they were two real but unrelated groups of people. *Khakian*, although not a traditional word, is derived from *khak,* which means "dust"/ "soil," translating into "those living on earth." But as a child, I remember, I had real problem comprehending what *aflakian* was all about. Part of the problem was that I knew exactly what *aflak* meant. It is the plural word for *falak*, meaning "firmament," the "sky." Thus, literally, *aflakian* means "people who live in the sky." But in those days, some 45 years ago, following the Western trends, we were not supposed to assume that there could be extra terrestrials living in the outer space. It was not as it is now that this notion is quite acceptable, at least for non-scientists–to assume that in the distant past, some aliens from other planets like the moon of Saturn, Titan, could have frequented the planet Earth. It should be noted that Saturn itself in not a planet entirely composed of solid matter; rather, it is in a gaseous state with a solid core. On the other hand, its moon, Titan, is the only moon in the solar system with a significant atmosphere.

So at the time of my childhood, the sky was the intangible realm of the intangible God and its intangible angels. But after close inspection of many religious scriptures or the literal works of many cultures, it is not always quite clear whether angels were merely spiritual beings or indeed some sort of living beings residing in the sky.

Since the emergence of a new economic system based on the Industrial Revolution of the West in the 18th century, many human endeavors from religion to science have become part of a bigger picture conveniently called progress. But progress as a process is fundamentally based on positive thinking and certainties. So let's make everything clear-cut. Let's make angels fictional and gods unattainable. Let's not confuse ourselves with ifs and buts. Let's make the human thoughts and concerns as standard as window frames and residential lots. Confused minds make good artists and thinkers, but let's guide people in striving for happy thoughts and strong bodies—that's what they really need, and that's probably what 99 percent of people want. That is indeed what "progress" has come to entail in our modern societies.

Sons of God

Although currently we are not culturally allowed to ponder about the nature of angels and other mysterious things, or to investigate them scientifically, we have managed to transfer their identities and personalities to certain type of humans, like saints and other holy people.

I have a small, one-foot tall statue made of lime of the Virgin Mary. The very same kind placed in a variety of sizes in the front yard of many Catholic houses. The one originally made in Italy in 1965. I have placed it next to a lampshade in my living room, and when, every evening at sunset, I turn on the lamp, from the angle the light shines on her beautiful face, it seems as if the Virgin Mary likes the light and smiles at me from the left corner of her lips. I call her Maryam *khanoom*, meaning "Lady Mary" in Farsi. My younger sister had bought it for me a few years ago for my birthday. Once, while my sister and I were walking in the Mission District, a working class and very colorful neighborhood in San Francisco, populated mostly by people of Central America, I saw her in a shop full of little saints. But I liked her the best; she was so serene.

My sister, who bought the statue for me, is tall, beautiful and very kind. She is a dentist and works in a non-profit clinic for immigrants in the Mission District. I call her the Queen of the Mission. When she walks in the Mission, almost everybody knows her there. Sometimes she says with much delight, "I cannot sit in a coffee shop there and read a book by myself;

sooner or later somebody comes to my table and asks, 'Doctora, doctora, can I sit and have a coffee with you?'" She really likes listening to their life histories and their problems.

She is also an artist. Her undergraduate major was in painting, and then she studied to become a dentist at the UCSF Dental School. She has the sensitive soul of an artist and that is why everybody loves her. Her dental assistants simply adore her. Although a Muslim, she is a devotee of Jesus as well, and one of her major installation artworks was greatly influenced by his suffering. In her latest work, using her dental hand drill, she has patiently engraved delicate arabesque or Quranic verses on the bones of a few dead people, which were painstakingly obtained, to acknowledge their dignities despite their silent struggles as ordinary human beings.

Jesus and Mary both are highly revered figures in Islam, in the Quran and among Iranian people alike. In the Quran, there is a *sureh* (section) named after Mary, and Jesus is one of the most frequently mentioned names in the Quran, along with Moses and Joseph of Canaan and Solomon. It is mentioned in the Quran that Jesus was created in a special way, in the same way that Adam, the first intelligent human in Middle Eastern folklore, was created. But Islam considers the Christian belief, Jesus as the Son of God, a blasphemy. In one of the most frequently recited *surehs* of the Quran, it states that God was not born and did not have any offspring. Every practicing Muslim recites this short *sureh* for a total of 17 times during the daily prayers. This is where Islam and Christianity have their major conflict.

Still this spiritual conflict does not keep Muslims from going to church occasionally. Whenever my mom visits us in San Francisco, the first thing she likes to do is to go to a church and light up some candles. Her mother's name, Maryam, is a very common name in Iran. It is the name of a pretty, small flower, and is also the same as Miriam, the Hebrew name for Mary. My grandmother was originally from Baku, Azerbaijan, formerly a state of the Soviet Union, and prior to that, on and off, a part of Iran. Although not an Azari Turk, like many people from those areas, she had blue eyes, light colored hairs and a very fair complexion. With mild manners and a simple lifestyle, she reminded almost everybody of a living saint, and in every sense she was a perfect Muslim. She used to come to our house every afternoon to check on us and before leaving, she would ask my mother whether she had said her afternoon prayer or not. My mother would typically respond by rushing to the porch adjacent to the living room, looking at the neighbor's wall, and coming back triumphantly, telling her that there was still sunlight on the wall, meaning that she had yet time to say her afternoon prayer a little later. That response was about the only way to get my mild-mannered grandma irritated. She would look at my mom, clearly disappointed, and would say, "Are you a sunlight worshiper or a God worshiper?"

I suppose that still in every Iranian, there is a little Zoroastrianism and Mithraism left, which were the Iranians' earlier religions before Islam. Sunshine, light and particularly fire are the symbols of Zoroastrians. The sun itself was the symbol of Mithraism, an earlier religion in Iran before Zoroastrianism, probably going back to the prehistoric times before the Indo-Iranian (Aryans) people split. *Mithra*, still a very popular name among Iranian girls, like several other deities, was mentioned in both Avesta and Rig Veda, the religious books of Zoroastrians and Hindus, respectively, pointing to the shared heritage of Iranians and Indians in very ancient times. *Mithra* was supposedly the deity whose birth date was on the 25th of December, or this date represented the day set aside for invoking its name. Due to the popularity of Mithraism among Roman soldiers in the early centuries AD, this date was later adopted as the birth date of Jesus Christ. Orthodox Christians still celebrate Jesus' birth date on the 8th of January.

One particularly interesting group of angels, with regard to the tangibility issue of angels, are "sons of God" who mated with "daughters of men" and fathered a new race of humans called Nephilim, as mentioned in the book of Genesis, who by some accounts looked like giant humans. Understandably, due to the sensitivity of this issue, there is quite a disagreement among religious scholars about what the Old Testament really meant by the "sons of God."

They were described in more detail in the book of Enoch and were called "Watchers." According to this book, they were 200 angels sent to earth simply to watch over the humans. But later their leaders, all with their own special names, defected and married humans and remained on earth, thus becoming known as fallen angels. Notwithstanding the foregoing, in some Christian traditions, "sons of God" is interpreted as meaning fully human and children of Seth (the third son of Adam), who married the daughters of Cain (the other son of Adam who killed his brother Abel and was banished to the east of Eden).

A Calendar Full of Angels

Zoroaster, an ancient Iranian prophet, by many accounts was the first prophet to introduce the concept of angels as watchers of humans, as in guardian angels. It is said that the revelation came to him at the age of thirty, when he went down to a river to get some water. While there, he saw a radiant figure who introduced himself to the astonished young priest. His light led Zoroaster to Ahura Mazda, the Wise Lord, and five other radiant figures.

The actual era in which Zoroaster lived and the place of his birth have been widely disputed, but according to many Zoroastrians he was born in Takht-e Soleyman in northwestern Iran some 5,600 years ago. By some other accounts, based on the similarities between the language and deities mentioned in his holy book, Avesta, and those in Rig Veda, the oldest religious text of Hindus, it is estimated to be about 3,500 years ago, somewhere in northeastern Iran.

It should be mentioned that many ancient religious books had been orally transmitted generation after generation through millenniums until, many centuries later, they were written down. One such book was Avesta, which about 1,500 to 1,700 years ago during the Sassanid Empire of Iran, was at last committed to the written medium. Zoroastrian priests were always concerned that there was no script rich enough to convey all the nuances embodied in the original language in which their sacred prayers and instructions were declared. Only after the development of a new alphabet with 16 vowels and 37 consonants, later known as Avestan, they did consent to have their sacred books be written down.

By some accounts Zoroastrianism started to flourish in Iran during the Achaemenid Empire, more than 2,500 years ago. Takht-e Jamshid, or as Greeks called it, Persepolis, the seat of the Achaemenid government, after so many centuries is still standing, reminding us of the brutality of time and the transient nature of mighty empires. It is a most majestic place with the ancient ruins of several imperial palaces, closely situated in a vast plain north of Shiraz, in southern Iran. The name ironically means the "throne of Jamshid," a mythical Iranian king, who probably lived several thousand years before the Achaemenid rulers. When standing amid all those towering columns and serenely designed black stone doorways, every Iranian indescribably feels a strong connection to an ancient heritage.

However, we Iranians don't have to travel far, wandering among old stones, glorifying some king of kings, to realize our antiquity. The ancient yet familiar names of the months of our national calendar, itself as old as the ruins of Persepolis, has its own story to tell. No longer recalling the significance of the names of the months of this calendar, still everyday by mentioning their enigmatic names, it makes us witnesses to a past that never ceases to disconnect or fade away and lay in ruins.

The current Iranian calendar, with some minor differences from its progenitor the Zoroastrian calendar, has the same origin and characteristics of the Zodiac calendar used in astrology, like the starting dates of the year and its months and the duration of each month. However, we do not use the Latin Zodiac names for our months. They have instead retained their own ancient names, each attributed to a specific deity, albeit with some modifications to their original Avestan pronunciations. Avestan is now an extinct language that was closely related to the more ancient form of Farsi and Sanskrit.

In this calendar, one month is dedicated to Ahura Mazda, the Creator or Supreme God as called by Zoroastrians. Six months are dedicated to *Amesha Spentas* (Holy Immortals), the "divine sparks" of the Creator, spiritual beings similar to archangels. It seems that these six deities are the personifications of the light figures Zoroaster envisioned as a young priest at the riverbank. The other five months of the calendar are dedicated to the most important of *yazatas* or, as we call them now in Farsi, *izads*.

Yazata is an Avestan word derived from the root word *yaz* meaning "to worship," "to honor." It means an "adorable being." There are many *yashts* (hymns/prayers) in Avesta in praise of different *yazatas*. Some of these *yazatas* also appear in Rig Veda, considered the oldest religious text of Hindus in India. It seems that the origin of these shared deities goes back to when Proto-Indo-Iranian people worshiped the same two groups of deities known as *ahuras* (*asuras* in Hindu) and *divs*.

I remember, back in my school years in Iran, there was a line in one of our national anthems saying, "The light of *izad* is always our guidance." We were told that *izad* in ancient times meant "God." *Izad* and another closely related word, *yazdan,* also meaning "God," although not used in our everyday spoken language, were and still are widely used in our literal writings as God. Later, I was very surprised to learn that *izad* really did not mean God, but something more like archangel. Actually, the concept of *izad* predated Zoroaster. Apparently, due to the persistence of the popularity of some prehistoric *ahuras* among people, they were kept as deities after the establishment of monotheism through Zoroastrianism. But they were demoted to *izads* to function below Ahura Mazda as its collaborators.

The advocacy of Ahura Mazda as the Supreme God by Zoroaster heralded a major advance in religious traditions, as well as new technical advances of humanity, at the time. Many deities in the earlier religions represented and protected natural elements and forces. With the advent of the Bronze Age, humans were no longer totally at the mercy of natural elements. The new superiority of Ahura Mazda as the Lord of Wisdom, upon placing humans under its direct protection, was to acknowledge the power of the mind and the new position of humans as beings capable of changing their conditions through thinking and wisdom.

Not only each month was dedicated to a deity, each day of the month was also dedicated to a deity. The first day of the month was dedicated to Ohrmazd, which was the original name of Ahura Mazda, the Creator. The name Ohrmazd probably was a reference to a primal mind, as in *ur*, meaning "primal," "early" and *mazd*, meaning "mind." In modern Farsi, it has morphed to Hormuz, known in the West mainly as the Strait of Hormuz, a 21-mile-wide bottleneck in the Persian Gulf, where currently about 40 percent of the world's oil passes everyday. Moreover, the tenth month of the Iranian calendar year, which is called *Day*, was dedicated to the invocation of the name of Ahura Mazda. I should mention that the Avestan name *Day* has completely lost its original meaning relating to God in modern

Farsi. Now, it is only defined as the tenth month of the Iranian calendar year. As a Proto-Indo-European word, *Day* is related to the word "deity" meaning "shinning."

In ancient times, when the name of a month and the name of a day coincided, it would signify an auspicious occasion worthy of celebration and invocation of the name of each specific deity. For example, Mehregan, an early autumn festival still celebrated in Iran, is the day when the seventh month and the sixteenth day of each month, both dedicated to *Mehr* (*Mithra* in Avestan), coincide. This *izad* represented "love," "kindness" and "light," and also the act of covenant between two entities, like the oath of friendship between two persons. According to Zoroastrians, invoking the name of deities is an act of worship in itself, commonly done by praising them in their specific prayers. Thus, naming the days and months of the year after them was an official measure to ensure that they would be remembered continuously.

The most mentioned and prominent of all *izads* in the Zoroastrian prayers was *Soroush* (*Srosh*, *Sraosha*), whom the seventeenth day of each month was later dedicated to. All ceremonies were started and ended by invoking its name. The name is derived from the root word *sru*, meaning "to hear," so *Soroush* means "listening"/ "obeying" the commands of Ahura Mazda. Due to this particular virtue, in Zoroastrianism he is regarded as the guardian of humanity for obeying the laws of Ahura Mazda called *asha*. According to Zoroaster's teachings, following *asha*, the natural order of all living and non-living things, was considered the highest protection humanity could have. The present-day word for *asha* in Farsi is *rast*, which is equivalent to the English word *right*. Like the meaning of right in English, *rast* means both "true" and "straight."

Soroush, among Zoroastrians, is considered the voice of conscience, as well as the intermediary between God and humans. He is the embodiment of the divine word. He is the protector of souls, and his aid is specially invoked at night. Its sacred bird is said to be the cock that heralds the coming of the new day. According to Zoroastrians, after death, the soul of the evildoer hovers, for three days and nights, over his/her head, weeping and saying, "Where shall I go and with whom shall I now take refuge?" While the soul of the good-doer, for three days and nights, sits patiently beside the grave where the body is laid. On the fourth day, at dawn, a demon comes and drags the soul of the damned to the *Chinvat Peretum*, the Bridge of Requiter, which the soul fails to pass and thus falls into hell. However, the soul of good-doer, accompanied by the Blessed *Soroush*, crosses over the Bridge into the next world.

Still a common name for Iranian girls, *Soroush* has also morphed into a popular name for magazines and newspapers or any publication with messages of hope and deliverance. After the demise of Zoroastrianism in Iran, like some previous pre-Zoroastrian *izads*, *Soroush* was able to retain its reverence as a highly regarded angel among Muslim Iranians. Due to its role as the messenger of the divine word, he was later identified with Jibrail (Gabriel), the angel of revelation in Islam. According to Islam, Jibrail, who is considered the same as the Holy Spirit (Ruholghods), is the archangel who delivered the word of God, the Quran, to the Prophet Mohammad.

Descent of *Faravashis*

Not every deity representing a specific day of month in the Iranian calendar fared as well as *Soroush* did. Nobody remembers anymore that each day of the month was devoted to a specific deity. There is no more invocation of their holy names. There is no more adoration. As humans need oxygen to sustain life, deities need remembrance and prayers to exist. A few of them have managed to remain in our everyday language—not as dei-tics, rather as common names for Iranian boys and girls. The most important of these forgotten deities is called *Bahman* (currently a common name for Iranian boys), after whom the second day of each month and also the eleventh month of the calendar year were named. He will be discussed in more detail in the next chapter (see p. 267). Also the names of a few others, who represented and protected natural objects, such as the sun, moon, sky and earth have survived as names only and not as deities. But for the rest, since nobody found any use for their names, they are forgotten and gone. All except one. Its name is *Farvardin*, which the nineteenth day of each month was dedicated to its remembrance and adoration. The first month of the Iranian year, which is also the first month of spring, is also named in its honor. Additionally, the Iranian New Year ceremony, called *Noruz,* is celebrated beginning on the first day of Farvardin, that is, the first month. With all these opportunities given to invoke its name, it must have been an important deity, or was it a deity?

Its Avestan name is *faravashi* (*fravashi*) and its Farsi name is *fareshteh* (*fereshteh*). It is not quite clear whether *Farvardin* and *faravashi* were just the same word expressed slightly differently in two previous Iranian languages, or whether each represents a different aspect of the same concept. Nevertheless, it conveys the original concept of guardian angel as it was acquired later, first by the Jewish people and then by Christians. But to define it just as a guardian angel would be a gross injustice to the vast spectrum of its other metaphysical attributes. Its origin also seems to predate Zoroastrianism. It is one of those perpetual concepts that has been part of the human psyche since we started to deliberate about the meaning of our existence on earth and beyond.

There is a fascinating story about the decent of *faravashis*. It is said that at the time Ahura Mazda finished the creation of both spiritual and material worlds, it also created, in the unseen world, the *faravashis* (roughly identified as souls) of all human beings. At the beginning, they were guarding the boundaries of heaven until Ahura Mazda figured out that he could employ them to win the battle between good and evil in the material world.

So he gave them an audience and asked them a fateful question. He asked them, "Which option suits you better? Would you rather stay here and be eternally preserved from the adversary of evil? Or would you prefer that I shape you into matter and send you to earth so you can live and dwell among lies and distortions, striving to overcome them and therefore destroy all evil forces once and for all, and thereafter live eternally with no enemies at all? And so they agreed to descend on earth in order to guard their human counterparts and stand by them to the end of their lives. Moreover, each one was to serve as an ideal, which the soul had to strive to emulate because while souls are supposed to be in the bondage of their physical bodies, *faravashis* are regarded to be free, guiding humans through their inner voices.

In Avesta it is mentioned that Ahura Mazda instructed Zoroaster to invoke *faravashis* for help whenever in danger. Furthermore, according to an ascetic who once lived in Tajikistan, people can achieve their dreams by telling them to their *faravashis*, and then they would notify the specific *izads* that they know of, through whom the dreams would reach one of the Holy Immortals and thereafter reach the throne of the Creator. By other accounts, *faravashis* act as guardian spirits who send out *urvans* (primal souls) into the material world to fight the battle of good versus evil. Then, on the morning of day four after death, each soul returns to its *faravashi* with the record of its deeds in the material world. In another variation, the reunion of all souls with their *faravashis* will happen at the end of time, when evil is finally defeated.

Of *Faravashis* and DNA Repair Mechanisms

The concept of *faravashi* becomes much more complicated when we learn that everything–from water to sky, to earth, to plants, to animals, to *izads* and even Ahura Mazda itself–has a *faravashi*. We are also reminded that each *faravashi* is tailor-made to its own counterpart. Could all these endless number of *faravashis* represent a parallel universe? Or do they just embody the energy of God in preserving the order in creation?

The notion that *faravashis* exist universally in all living beings, but are uniquely defined within each of them, as well as the Zoroastrians' emphasis on the preservation of *asha* (cosmic and natural order) and the employment of *faravashis* in fighting earthy distortions (as described in the Farvardin yasht of Avesta), reminds me of something else. As a geneticist, it sounds to me like DNA itself; and it also reminds me of one of the most common genetic processes occurring in almost all organisms, that is, DNA replication and its fateful copying mechanisms.

If we could imagine DNA as a genetic deity, its characteristics are quite similar to those of *faravashis*. For example, the following descriptions variously stated for *faravashis* might as well be attributed to DNA: They keep everything in order, including the course of development of embryos in the womb. If it was not for their protection, all animals, plants and people could not continue to exist because the wicked lies (deleterious mutations/cancer?) would destroy them all. The body of a human, with its peculiar physical, mental, moral and spiritual capacities, is shaped and formed after the model, which each particular *faravashi* presents.

Similarly, DNA replication and DNA repair mechanisms proceed with very high fidelity to ensure smooth molecular operations in somatic cells during each organism's lifetime, and also to ensure the intact transfer of genetic materials to the next generation through germ cells. This high fidelity, although not absolute, is achieved through a proofreading mechanism involving several proteins and enzymes, which protect DNA by removing the misplaced nucleotides (mutations) and inserting the accurate ones. However, it is a constant battle for DNA to keep its contents undisturbed as much as possible and free from point mutations.

Needless to say, this comparison was not made to imply that *faravashis* are indeed DNA or any kind of material substance. After all, not only living beings but also all beings, according to Zoroastrianism, have their own *faravashis*, which does not apply to the nature of DNA. But the reason I brought it up here was to point out the similarities between scientific and religious thinking—one supposedly based on reasoning and the other based on revelation. One was figured out in the 20th century and the other was laid down four to five thousand years ago, both arriving at the same conclusion on how different aspects of nature try to preserve themselves (see also p. 264).
Another name closely related to *faravashi* is *Farvahar* in Avestan, or *Farohar* in modern Farsi, which is believed to mean the depiction of *faravashi*. It is a very familiar yet enigmatic figure, seen in the ancient ruins of Persepolis in Iran, which later became the symbol of Zoroastrianism worldwide. *Farvahar* will be discussed in detail in the next chapter (see p. 281).

Languishing Departed Souls

In Zoroastrianism, as mentioned above, *faravashi* is closely associated with the human soul. Sometimes it is portrayed as external soul, pre-existing the body, but working with the human's vital spirit. And some other times, it is said to represent the divine and uncorrupted part of the soul kept in the bondage of the body. In both cases it protects a living being, but *faravashi* has also been described as the immaculate part of the soul of departed people, or more specifically as the departed soul of righteous and pious people.

It is still customary among Zoroastrians, during their special ceremonies, to devote sometime for the invocation of the *faravashis* of some specific departed souls. It is also considered a religious act to recite the Farvardin yasht (prayer) in memory of their departed souls on their death anniversaries, on the nineteenth day of each month of the Zoroastrian calendar, and also during the days of Farvardegan. This is a festival celebrating the return of departed souls to their former homes at the end of each year. They were believed to arrive at nighttime and then depart the following day at sunset.

The origin of this particular celebration seems to have predated Zoroaster, when, during the earlier times, departed souls were just departed souls. When there was no concept of guardian angels or *faravashis* protecting humans during their lifetimes from evil acts. And when there was no hint of salvation of the human soul through her/his choice to side with goodness in the cosmic battle between *nicky* (kindness) and *bady* (to do bad).

According to pre-Zoroastrian ancient beliefs, probably going back 5,000 years, after death the *urvan* (the same as *ravan* in modern Farsi), meaning "primal soul"/ "soul," would hang around for three days and then with no hope of returning to life, would slip through its grave into the underworld of the dead. Then understandably, especially during the first week, it would feel very lonely, rejected and desperately in need of attention from its living relatives. The family support would come through special prayers and offerings, such as giving food to the poor with the special request for praying for the departed soul to rest in peace. It was believed that for one year the departed soul would be still longing to come back and would feel like a stranger among other departed souls, until little by little it would get used to its new conditions.

Ironically, even though the majority of we Iranians are now Muslims, our feelings about dead people still go back to our ancient non-Islamic beliefs. I remember that when for the first time my dead grandmother came to me in my dreams and talked about wishing to go to Mecca for the pilgrimage, my mother first gave food offerings to some poor people, and then arranged for a person to go to Mecca in her place. All my four grandparents departed a long time ago. But only two of them kept coming to me in my dreams for a long time. These were my two grandmothers. The one who wanted to go to Mecca used to come to me in my dreams rather infrequently. She always looked the same as I remembered her when she was alive.

However, the story was very different for my other grandmother who, during her lifetime, had numerous friends and traveled to many places, including Mecca. She was constantly in my dreams and always extremely sad. There is no word in this world for me to describe how sad and morbid she looked in my dreams. Sometimes, I would see her sitting alone in a large and very lighted underground prison, always with her head down and eyes closed with a face, as we say it in Farsi, burdened with a mountain of sorrow. Still, I would consider those as my good dreams about her. On many other occasions, she appeared in my dreams as a figure sitting dead among living people, emanating a very heavy aura of death all around her with an unbelievable sadness. But most times, I would see her sitting in her open grave, her body rotting, half dead, half alive, and unbearably sad.

The only time she appeared normal and in a good mood in my dream was when she and my other grandmother were in a large gathering of people together. I do not recall at all whether it was my last dream of them or not. But since last year, I do not remember dreaming about them anymore. So I wonder, what happened? Is it me who does not dream about them anymore or maybe I do not remember my dreams as well as I did before? Or is it them who do not come to my dreams anymore? Then, I thought maybe the soul of dead people is also regulated by time, as if they have a finite time to be around and then after some thirty years they disappear.

As a scientist, it was just like an interesting working hypothesis to me and I really did not believe it. Until a few months later when I read an interesting article about the Iranian ancient beliefs about the departed souls. It made me suddenly very sad, as if my grandmothers were dead again, or as if they were really dead this time. It said that, according to some ancient beliefs, offerings for the departed souls would be made for up to thirty years. I repeated to myself: "thirty years," it definitely sounded familiar to me. So who knows? Maybe thirty years was not pure superstition and was based on some observations that people made in ancient times.

My sadness was not because I may not see them again in my dreams; rather I suddenly felt that I had no idea where they, or other departed souls for that matter, are now. The house of cards tumbled down for me in a matter of seconds. All those fancy indoctrinations about souls that we hear, read and see in movies here and there suddenly disappeared. The ones that imply after death our souls go to a peaceful place beyond any earthly torments. Suddenly, I was not sure of anything.

My mind traveled 5,000 years back and I felt solace in the sincerity of ancient beliefs about departed souls, when there was no option for the soul to go up, only to go down. That was apparently the same time that people started to invent better tools to work up the land more efficiently and to settle down and have a roof over their heads and live a semi pastoral life. No more wandering around for the next meal.

But settlement had its own problems. There were always people who wanted to grab other people's lands. Likewise, there were natural disasters, which ruined people's cultivated lands. Thus, warriors and priests started to gain prominence among other people, the first group to protect the land, and the latter to mitigate with gods to avert disasters and get more from the land. The priests also originally functioned as teachers. So in no time three classes of people were established and most people, those who were neither warriors nor priests, were obviously not happy about this new social order.

Then, came the promise of heaven for everybody. Not only for the souls of brave warriors and holy priests, but also for the souls of farmers and laborers, only if they could prove to be pious enough to deserve such a coveted heavenly place. Thus, through the ages people have converted to new religions again and again with the promise of equality. But not everybody is affluent enough or that naive to pick up a new religion based on the equality of souls in heaven. Most people, first and foremost, want equality here and now.

Who Was the First Man of Manicheans?

Creation Myth of Mani

A Forgotten Prophet

Adam looked around and wept, expressing his deepest anguish while uttering, "Cursed be the creator of my body and he who bound my soul and those who have made me their slave." The ancient Iranian prophet, Mani, wrote down this sentence about Adam's lamentation, seventeen centuries ago, as part of an expansive and most amazing creation myth. The moment I read this passage on an Internet web site on October 7, 2004, I was so taken with his universal plea that a vague thought of writing a book about this subject entered my mind. The mention of an imprisoned divine soul, all of a sudden, brought back many distant memories of a cultural heritage that I had disregarded so carelessly in my youth, in favor of some modern rituals.

But it took me another six months, after watching a prolonged TV drama series (see p. 103), to find enough courage to convince myself, as a geneticist, that it might be the time to write a book for the general public about a subject as irrelevant to our current scientific agenda as this one. Something like initiating an open dialog with typical non-scientist people about what genetics is, and more importantly, what genetics could be when contemplated outside the boundaries of its current established format. If nothing else, I thought, it would be a perfect forum to see whether or not I was ready to untangle myself from all those superfluous scientific ideas that once sounded so logical and endearing to me.

I was entranced with the Prophet Mani's account of the human creation, not as a romantic fairytale, but rather as a viable concept, something to ponder about. And intriguingly enough, it soon became a device to test myself with, like holding a mirror to my soul to look at my real thoughts, those engraved deeply, not just some ornamental ones hanging around to fall down sooner or later with the passage of time.

Indeed, the concept of soul or a part of it as a divine spark, so passionately expressed by Mani, has a long and continuous history in Iranian religious philosophy. It reminded me right away of the ancient Iranian concept of *farr-e izadi* (see p. 278), of the longing for unification with the Supreme Being in Sufism as expressed so divinely in the mystical poetry of Hafez and Rumi, and of Sohrawardi's complex ideas as revealed in his 12th century School of Illumination (see p. 294), a religious order quite similar to Sufism, which is a multi-cultural Islamic esoteric and mystical tradition.

The Prophet Mani was born around 210 AD, in the ancient city of Ecbatana (modern-day Hamedan) in western Iran, from an aristocrat mother and a puritan reformer father. His father soon left his birthplace and took his young son with him to Babylon (in central Iraq), then part of Iran and a center of ancient religious thoughts. By some accounts while there he joined Mandaeans, a group of people devoted to a pious way of life following their ancient traditions (see p. 218).

Mani apparently received his first revelation at the young age of twelve from a spirit or a voice inside him that he later called his twin, his double, his protective angel, or his "divine self." The equivalent word that we use in Farsi is *hamzad*, literally meaning "something or someone who is born with you." According to his teachings, this inner voice presented him with the divine knowledge, which later was developed into Manichean religion. But the voice asked him not to reveal his prophecy for another twelve years, during which Mani had enough time to be trained as a painter and became a master in this field. His main religious text, called *Arjang*, meaning "precious," described his elaborate cosmology of creation and was filled with masterfully painted pictures of the universe. For this reason, he is still better known among Iranians as Mani The Painter.

He was known among his followers as the messenger of light. His name, Mani, although currently a common name among Iranian males, was not his real name. It was his title and a term of respect. The word *mani* more likely have originated from *mana*, a word used among Mandaeans (see p. 218) meaning "light spirit." Mani lived long enough to travel and spread his teachings from Babylon to India, and then to Turkestan (now called Central Asia). Later, Manichean religion spread from India To Tibet and from Central Asia to China, where in one of its northern states it became an official state religion. In the West, it had considerable following in many parts of Europe and North Africa. Additionally in Europe, Manichean concepts greatly influenced several Christian Gnostic orders that later were branded as heretics by the Christian Church.

In the end, when Mani was in his early sixties, after a lifetime of on-again, off-again support from a number of the Sassanid kings of Iran, the verdict went completely against him. At the urging of the Zoroastrian high priests of the royal court, the Manichean heresy, as it was called later, was brutally rooted out in Iran, once and for all. Mani himself, by some accounts, after years in prison, was eventually crucified, his corpse flayed, the skin stuffed and hung up at the city gate. The horrific punishment was meant to show that it would not pay to go against the mighty traditions.

Sadly enough, his teachings were essentially based on Zoroastrian principles with some additional revelatory insights foretelling an elaborate and detailed version of the struggle between the forces of light versus darkness. It seems that his doom was eventually sealed not only because he went against traditions, but rather against the powerful Zoroastrian religious establishment of the time.

The ancient land of prophets, whether officially Zoroastrian or officially Islamic, in the course of its long history, was never kind to its new prophets. In traditional Iran, one is born with her/his religion, whatever it is, and is supposed to die with it. Like an old curse, religious conversion equates with abandoning one's own heritage.

After Mani's demise, although Manicheaism rapidly died out in Iran through persecution, and the subsequent escape of its followers to Turkestan (Central Asia), it remained a vibrant religion for one thousand years outside the Iranian borders with a vast number of followers in Africa, Asia, and Europe. Then, it faded out. Part of the universal appeal of his teachings was because they included many ideas from several other religious traditions, such as Zoroastrianism, Christianity and Buddhism, as well as Babylon's folklore. Manicheaism is considered by many theologians as one of the most influential religions of the past, in part because of its great influence on early Christianity.

Saint Augustine, who is one of the fathers of the Christian Church, was a devoted Manichean for ten years. He was born in the 4th century AD in Algeria, in North Africa, and was most likely of Berber descent. His mother, Saint Monica, was raised in an early Christian household in Algeria, and brought him up as a Christian, but later in his youth he converted to Manicheaism, and then he converted back to Christianity.

The First Man, the Forerunner of Adam

The drama of Manicheaism evolved, as foretold by Zoroastrianism, around the eternal conflict between the spiritual world of light and the material world of darkness. Apparently, before the existence of heaven and earth, the forces of light and darkness both existed from time immemorial in two parallel universes, with the distinction that the light realm remained forever uncorrupted, whereas its primal copy, the dark realm, while succumbing to frequent distortions, went through ever-lasting change. But Mani brought to life such an elaborate and detailed description of the ongoing events, with a cast of so many colorful personalities, that it is very hard to consider them purely imaginative or symbolic. It somehow brings to mind the images of the alien-like inhabitants of some faraway celestial places frequenting the planet Earth, for some unknown reasons, with their shining and light-filled starships.

The following is a very brief description of Mani's amazing cosmology and creation myth occurring in three stages. The emphasis here is only on the events leading to the creation of a mythical character very similar to Adam.

One day, a menacing monster from the underworld invaded the realm of light. In the first stage of creation, from the realm of light, *Nasa Qadmaya* (meaning the "First Man" or an "ancient person" in Aramaic), dressed in an armor of light was sent to repel the monster of darkness and his demonic lieutenants. Soon, he was defeated and was dragged down to the realm of darkness by its demonic forces. After recovering his sense of reason, he made a call to the realm of light for help. In response, in the second stage of creation, a new agency was created in the realm of light and was sent to rescue the First Man. His eaten and beaten body was then raised to the realm of light, but his shining armor was left behind, which was soon devoured by the jubilant demons (*divs*?) of darkness. Unable to retrieve completely the light armor of the First Man, the regents started to make the world, as we know it, in the realm of darkness during the second stage.

A new agency, thereafter, was created again in the realm of light and was sent to the realm of darkness. But it seems the negotiations carried out between the messenger of the realm of light and the demonic sons and daughters of darkness ended with little success and finally from *Naimrael* and *Ashakluna*, a female and male demon (*div*?), respectively, two children with characteristics very similar to Adam and Eve were born. Since in the Manichean literature of the West, they are currently called Adam and Eve, here I would rather call them the same.

In the body of Adam a great number of light particles, as originally devoured by the demons, were imprisoned, and he himself was a captive of the force of darkness, without any knowledge of his past. The realm of light, still concerned with the light imprisoned in the soul of Adam, sent a savior for Adam to let him know of his divine origin. The savior also told him all about the suffering that the divine spark imprisoned in him has endured during its fall and long captivity in the realm of darkness. Adam stood up, looked around and wept while uttering, "Cursed be the creator of my body and he who bound my soul and those who have made me their slave."

According to Manichean teachings, the ultimate aim of humanity was to release the light of essence from the pollution of the body. Manicheans believed that eventually, through the acts of pious people, all light particles would gather together one day, free from the bondage of matter. To help bring that day closer, they advocated a way of life dedicated to avoidance of evil thoughts, evil words and evil deeds, as well as a restrict vegetarian diet and great respect for nature and its elements, as originally advocated by the Zoroastrian codes of conduct. Additionally, they mandated a much stricter lifestyle for those among themselves who could undertake it. For example, devote

Manicheans were not even supposed to prepare their own food because picking fruits and vegetables were deemed intrusive to nature. Moreover, since the propagation of the human race meant a continuation of imprisonment of the light source, marriage was considered a despicable act, and maternity was viewed as a sin.

The Central Dogma of Molecular Biology and Mani's Cosmology

Going back to Mani's cosmology, at first it seemed like a very imaginative scenario to me. Then, when I started to contemplate its every little detail, his insight into the nature of things started to look more and more revelatory than imaginary. For example, at the first and second stages, he talked about the creation of three figures that was meant to represent unity, and at both stages, only the third figure would go and do the actual work. So during the first stage, when the realm of light was invaded by the forces of darkness, the Supreme Being of this realm, *Abba de-rabbutha*, meaning the "Father of Greatness" says, "I will go and fight with him myself," which meant that he right away emanated the "Mother of Life," and she emanated the "First Man," who was actually the one who was sent down to fight the forces of darkness.

The first question that comes to mind here is: why did the Father of Greatness, who originally said he would go and fight himself, not do it and instead created two other figures to carry out his plan? Anybody who has taken Biology 101 in college and still remembers it would be able to identify with the nature of this holy trinity. The way it works is indeed very similar to the central dogma of molecular biology. DNA is transcribed into RNA and RNA is translated into protein, which actually does the real work in the cell, as a structural protein, an enzyme or a hormone. In other words, generally speaking, any stretch of DNA that corresponds to any specific gene manifests itself first in the form of an intermediary named messenger RNA, which consequently makes a specific protein representing that gene. The messenger RNA involved here in this process is not an exact copy of that particular piece of DNA, but it represents only that piece.

The reason DNA functions this way is simply because it is the original copy and cannot afford to get corrupted. That is why, in eukaryotes, it is housed in a special compartment called nucleus to keep it away from all the chemical reactions happening in its surroundings, the cytoplasm of the cell. Proteins, on the other hand, are made outside the nucleus in the cytoplasm through an elaborate cellular mechanism. Trinity in unity, as seen in Mani's creation myth and also in molecular biology, seems to be one of nature's primary codes of conduct as expressed in the simple form "thought, word, action." But figuring out this natural order is currently attributed, in the West, only to scientists. However, it is not unreasonable to assume that a few non-scientist people like prophets and sages could also have the same or even higher levels of understanding of the laws of nature.

Ironically, the central dogma of molecular biology, as mentioned above, is the product of many million years of evolutionary processes, supposedly occurring by pure random mutations. So in light of this, how can we reconcile the following anomaly? How could a randomly occurred outcome like the central dogma of molecular biology and the concept of trinity in unity, a product of human faculty, and expressed commonly in many religions, be so similar? One explanation could be that the minds of some people like those of prophets and sages might be somehow aware of the body's chemical machineries, possibly through some inner-body relay systems.

Ancient Deities or Visiting Outer Space Aliens

Adam and Eve and their counterparts are at the center stage of the creation myth of many ancient cultures. But as we have seen in the previous section, in the creation myth of Mani, before Adam and Eve there was the First Man, whose arrival on earth initiated a chain of events ultimately leading to the appearance of Adam and Eve as the progenitors of the humankind. This makes the First Man even more mysterious than Adam and Eve. Who was he? Was he a real cosmic being or an imaginary figure?

From the literal interpretation of Mani's creation myth, it seems that the First Man came from outer space. So was he an alien, a visiting alien? Of course, there is no reason to take Mani's words and his cosmology seriously. But Mani was not the only sage talking about these perplexing issues. In the narratives of many ancient cultures, there was constant mentioning of the encounters with mysterious shining living beings, which were apparently not from the planet Earth. Ironically, many of these mystifying figures later morphed into objects of worship as various deities in ancient times, as if the deities were the collective memories of some unexplained encounters with superior non-human beings that happened in a very distant past.

For example, the Prophet Zoroaster claimed that he had encountered seven shining figures who talked to him at a riverbank when he was 30 years old. So his claim and other similar claims could be regarded as either real or unreal. Some people could argue that not only some humans have active imaginations; they would also like to entertain themselves with bizarre stories. On the other hand, there is no reason to think that they could not be real. Just because something does not fit into our contemporary worldview, it does not mean that it should be imaginary and unreal. If, for the sake of argument, we assume that these stories were real, they could have been describing tangible or intangible events. For lack of better words, what I mean by tangible events are those events that everybody with eyes or ears could witness. On the other hand, intangible real events would be the ones that, while not the products of imagination, only a few people would be able to perceive. As if these events would momentarily flash from a hypothetical "parallel universe," passing through time either concurrently with our universe or out of sync with our time dimension, either belonging to the past or the future.

The First Earthly Regent

Going back to Mani's First Man, it should be mentioned that the conception of the First Man as a living being, who was the forerunner of Adam and Eve of the Old Testament and other Middle Eastern traditions, is uniquely an Iranian notion. Interestingly, there was a similar figure to the First Man before Mani's time in some earlier Iranian cosmologies, first in Mazdian cosmology and later in Zoroastrian cosmology. However, in the earlier cosmologies, although he was similarly the progenitor of the first pair of male and female humans, there was no indication that, like the First Man, he had a cosmic origin; rather he was the earthly regent of a cosmic figure. His name was Kayumarth (Kewmarth/Kiamarth). *Kay/kayu* means "life" and *marth* means "mortal," so together it means "mortal life" or "mortal living being."

I still remember the first time I read about Kayumarth, in an Iranian encyclopedia that I had had since my teenage years. It said, "He was called the first king and by many accounts the first *bashar*." Interestingly, in Iranian mythology, he is usually called the first *bashar* or first *ensan*, but not Adam. In our current Farsi usage, *bashar*, *ensan* and Adam all are generally used interchangeably to mean "human." As will be described below, not only did I find the Kayumarth character ambiguous, but also with the Western mindset that I had then, some thirty-seven years ago, his genetic contribution to the humankind seemed utterly incomprehensible and meaningless.

In the Mazdian cosmology, Kayumarth was created after the creation of plants and animals. He was assigned by *Vohu Manah* (good mind or good nature) to keep the realm of good nature permanently out of the reach of *Ako Manah* (evil mind or evil nature), and with the help of celestial agents of order and inspiration transformed it into a state of perfection and consequently immortality, never to be corrupted again with the forces of darkness. But *Ako Manah* managed to break the sky ceiling and entered the realm of good nature destroying everything there including Kayumarth and the bull that had lived with him for 3,000 years. Then, after his massive destruction, when *Ako Manah* tried to leave the realm of good nature, he could not pass through it because the hole that he had made in the sky upon his entry had already been patched up by the agent of order in the realm of good nature. So *Ako Manah* became trapped in this world, which made the ideals of perfection and immortality and therefore, the emergence of a blissful realm, unattainable.

Of Cellular Mechanisms and the Mythical Ideals of a Perfect World

When I read about *Vohu Manah*'s intension in creating a blissful realm first through achieving a state of perfection and then immortality, it reminded me again of the biological realm of DNA. However, in evolutionary terms this concept does not mean anything because evolution through endless random mutations does not foresee any direction, everything happens by pure chance. As commonly expressed, evolution has past but no future, unless we start to think about other mechanisms for evolution, like evolution through canalization (see p. 141). On the other hand, if we do not consider the evolution of a living being as a whole, rather only its DNA part, then we can see, with no doubt, a genetic counterpart in the myth of *Vohu Manah*. That is, if DNA had its own mind and wanted to live in a blissful realm, then for reaching perfection and immortality, it would have to devise some specific mechanisms.

The function of these mechanisms, needless to say, has to be the preservation of the exact building blocks of DNA sequence, so random mutations would not be able to alter it permanently. Indeed, DNA is already equipped with such mechanisms. However, these are not error-free processes and that is what allows the evolution to happen. Currently, there are two known mechanisms that enable DNA to deal with random mutations with extreme efficiency. One is the DNA proofreading system that works during DNA replication (copying). The other one is an extensive DNA repair system in operation to correct the mutations caused by UV radiation and also by exposure to environmental carcinogens (cancer causing agents). So here we have again encountered another example in which an ancient myth shows the same principles as two well-known biological processes (see also p. 245). Is it accidental or is our human mind unconsciously aware of our universe within?

A Most Bizarre Creation

Going back to the Kayumarth myth, later when Zoroastrian cosmology replaced and reformed the Mazdian cosmology, the Kayumarth story was modified to some extent and Ahriman (a new version of *Ako Mana*, the same as Satan) did not kill him. Rather, Ahriman made him mortal, which made the Supreme

God, Ahura Mazda, very angry. Therefore, when death approached Kayumarth, his seed was taken to the moon, got purified by sunrays and returned to earth.

Again here, if we take these events literally, some aliens must have been involved in performing these rather strange tasks. According to this version of the creation myth, from his seed a rhubarb-like plant grew with two joined stems, one called *Mashiya*, and the other, *Mashiyana*, who after turning into humans became the mother and father of the humankind. It should be mentioned that although presently Kayumarth is a common name for Iranian boys, the names *Mashiyana* and *Mashiya* and even their stories are completely unfamiliar to most Iranians.

The concept of a primal pair of humans with opposite sex and joined together, although quite strange, has survived in the texts of many Gnostics traditions. For example, in the book *Apocalypse of Adam*, Adam on his deathbed disclosed to his son Seth that originally he, Adam, and Seth's mother, Eve, were joined together until God got angry with them and separated them from each other. He also stated that God got angry with them because the outcome from the creation of Adam and Eve proved to be better than his own creation (God's), implying that the god he was referring to was not the Supreme God, rather one emanated from another being. It should be mentioned that when this book was written around the early centuries AD, the general belief among Christians was that there existed a higher God above the God of the Old Testament.

The Gnostic notion of the necessity for the presence of both female and male elements in the creation process has also its counterpart in genetics. The reason that the rate of successful animal cloning is so small, about 1 in 300 trials, is a powerful testament to a recently developed genetic principle called genetic imprinting. For embryonic development to continue properly, it needs both maternal and paternal genes coming directly from the egg and sperm. Just to have two sets of genes coming from somatic cells (typical cells in the body), as the definition of cloning implies, is not always enough. Genetic imprinting is a process in which some genes get differential treatments in males and females during gametogenesis, that is, the process of producing the eggs and sperm. Therefore, for many genes, a copy from each parent is necessary for embryonic development to proceed accurately. Interestingly, this well-studied phenomenon in genetics is not related to the function of the genes on the X and Y sex chromosomes.

Likewise, the concept of a rhubarb-like plant as the container of the first human male and female, although sounds childish to say the least, may signify an artificial biological process, such as cloning or a more advanced and unrelated procedure involving the genetic manipulation of the living beings.

The Wise Lord

Going back to the First Man of Mani's cosmology, as pointed out before, a small fraction of the light armor he left behind ended up inside Adam, endowing him with a higher consciousness. Likewise, Kayumarth of Zoroastrian cosmology left behind his seed, which was used to create the first pair of male and female humans. There is yet another intriguing character in these stories whose legacy is also defined by his intellect and wisdom. As already mentioned above, he was *Vohu Manah* of Mazdian cosmology.

In ancient times, *Vohu* meant "good" and *manah* meant "thought"/ "mind"/ "knowledge" (see also p. 255). In modern Farsi, *vohu* has morphed into *bah* with the same meaning, whereas *manah* in the form of *mana* means "something that remains for a long time." Also *man*, which means "I" in modern Farsi, might be a derivative of *mana*. If so, it would be interesting that the word *mind* had changed into "I," as if the human existence was defined by the mind. *Vohu Manah* sounds similar to the English word *human*, which is not unexpected as English and Farsi are in the same family of languages.

There is additionally a quite similar word to *human* in Farsi called *hooman*, which, however, does not mean "human," but is rather a common name for Iranian boys. A newer version of *Vohu Manah* in Farsi is *Bahman*, which is also a common name for Iranian boys. It means *nik nahad* in modern Farsi. *Nik*, like *vohu* and *bah* simply means "good," whereas *nahad* has rather a complex meaning. It does not directly mean "mind." It means something like the "nature," "base," or "essence of a human being." *Nahad* also means the "act of putting or setting something." So, *nik nahad* might also mean "someone who has received a good thing" or probably a "good mind" as the ancient meaning of *Vohu Manah* implied.

So who was *Vohu Manah* in the pantheon of Mazdian deities? In Mazdian cosmology, at the beginning there was only void, which later gave rise to *manah* (mind). *Manah* created *Vohu Manah* and *Ako Manah* with opposing natures. *Vohu Manah* then, as the Mazdian lord, created successively his five celestial agents to assist Kayumarth transform the transient dominion of good nature into the eternal realm of the blissful. Later, in the Zoroastrian religion, Ahura Mazda, the Supreme God, and meaning the "Lord of Wisdom" was given both the name and function of *Vohu Manah*.

At the same time, *Vohu Manah* was turned into *Vohuman* and demoted to a lower deity. He, along with his five agents, constituted the *Amesha Spentas* (Holy Immortals) of Zoroastrianism. In this reformed order, Ahura Mazda delegated the affairs of living beings on earth to these six deities. As expected, *Vohuman* was the highest ranking among these six deities and always occupied the second place, next to Ahura Mazda, as evident from his ranking in the Zoroastrian calendar (the same as the current Iranian calendar, see p. 239). In this calendar, the eleventh month, Bahman, was dedicated to *Vohuman* right after the tenth month, Day, which was dedicated to Ahura Mazda. Moreover, as the first day of each month was dedicated to the invocation of Ohrmazd's name (another version of Ahura Mazda's name, probably referring to the primal mind or wisdom), the second day was assigned to *Vohuman* who is currently called *Bahman*.

Vohuman as a Holy Immortal became the personification of intellect, both innate and learned, as proclaimed in the following Zoroastrian prayer used for the invocation of his name "To *Vohuman*, to peace, whose breath is friendly, and who is more powerful to destroy than all other creatures, to the heavenly wisdom, made by Mazda, and to the wisdom acquired through the ear, made by Mazda." As the angel of wisdom, *Vohuman*'s task was to improve the mind of humans and also calm their anger and keep them sane while interacting with each other. Additionally, he was in charge of Zoroastrian astrology and medicine. So it seems in the same way that Ahura Mazda delegated its wisdom to *Vohuman* in the Zoroastrian cosmology; *Vohu Manah,* in the earlier Mazdian cosmology, had assigned Kayumarth to personify him and carry out his celestial plans on earth.

It should be mentioned that although the ancient Iranian cosmologies have long lost their relevance in Iranian culture altogether, the ancient concept of wisdom as human's best virtue has survived Iran's many religious transformations. The Iranian poet, Ferdowsi, in his epic poetry book, *Shahnameh*, uses so many times the word *Kherad* meaning "wisdom" as if it were the only human mental attribute.

Going back to the First Man of Mani, his "perfect man," the archetypal man, he was so revered among some Gnostic traditions that it is said that they foretold that the secret name of God was the real name, now lost, of the First Man. It should be noted that even today, among Sufi mystics, contemplating about the secret name of God is considered a major approach in getting close to God. Moreover, according to Iranian folklore, invoking the secret name of God was enough to make *divs* and demons disappear.

After the demise of Manicheans, however, the First Man, who St. Paul called the manifestation of God before the beginning of the worlds, soon lost his reverence among Iranians. In mainstream Islam of Iran, he was overshadowed by Adam who has been mentioned several times in the Quran, including how God created him from soil. Ironically, Iranians do not name their sons after Adam, but this name is used as a common word meaning "human," regardless of gender. Nevertheless, Kayumarth, a similar character to the First Man, has remained a common name, although people do not have much understanding of what the name is really representing.

In addition to Adam, we have another general name for "human." It is called *ensan*. If one asks any Farsi-speaking person whether there is any difference in the meaning of these two words, the answer most probably would be "no." But there is a subtle difference in the usage of these two words. In Farsi, we usually say that such and such a person is an "Adam" when we want to emphasize his/her civilized manners. On the other hand, when we like to emphasize his/her kindheartedness we call her/him an *ensan*. Those who have read the Quran may or may not recall, that there could be another distinction. One *sureh*, which addresses the humankind, briefly mentions without any further deliberation "*jin* and *ens* who we created before you." *Ens* here is the same as *ensan*. There was also the notion of *ensan-e kolli* in some medieval esoteric Iranian traditions. He was identified with the "universal man" who was regarded, as the heavenly counterpart of each human, somewhat similar to the character of Mani's First Man.

The first time I became aware of a possible difference between Adam and *ensan* was two years ago through the poetry of Hafez. Then as part of my work, I used to drive to San Jose, located about 50 miles south of San Francisco. To make good use of the long one-hour drive, I decided to listen to a CD of Hafez' poetry that my sister had brought for me from Iran several years ago. After listening to it a few times in my car, I was so taken with his revelatory insight and poetic splendor that to understand his poetry better, I decided to listen only to this CD during my weekly trip to San Jose, for as long as I could take it and not lose my interest in it, which amazingly took one whole year to happen.

So one day while listening to one of his poetry pieces, I noticed one particular verse that really intrigued me. With my utmost confusion, he talked about Adam and *ensan* as if they were two different individuals. Also he mentioned a third person whom he only referred to as "your/you." Referring to the scene of the creation of Adam by God in the Quran, in which God asked all angels including Satan to prostrate in front of Adam, Hafez says, "God in the prostration of Adam meant your adoration, because in your virtues, God found something above that of *ensan's*."

The elucidation of poetry of great Sufi poets like Hafez and Rumi is not an easy task. Since Sufis, devout Muslims themselves, did not agree with the mainstream interpretation of the Quran, to evade the Islamic authorities of the time, they usually spoke in an esoteric language that only they could understand. In Iran, there are scholars called *Hafezshenas,* meaning "Hafezologist," whose main expertise is the interpretation of the poetry of Hafez, and many times they do not agree among themselves about what Hafez is really talking about.

The Giant Man of Mandaeans

In another Middle Eastern Gnostic tradition practiced by Mandaeans, a group of people (see also p. 218) living in southern Iran and Iraq, there is a human-like giant called *Ur* who preceded Adam in the Mandaean creation myth. *Ur,* which means "primal", was also the name of an ancient Sumerian city on the Euphrates River in present-day southern Iraq, which is also said to be the birthplace of the Prophet Ibrahim (Abraham).

According to the creation myth of Mandaeans, *Alaha*, the Supreme God, brought 360 *melkis* (*malik* in Arabic) or regents into existence by pronouncing their names. Then *Alaha* delegated the creation and government of the material world to them. One of them, *Hibil Ziva*, created our earthly world, as well as a "world of light," *Mshuni Kushta*.

In this narrative, there is also a woman named *Ruhayya* (*Ruha*, see also p. 220) who lived in the lowest underworld of darkness. When she was pregnant with the son of one of the *melkis* named *Ashdum*, *Hibil Ziva* brought her to the upper world of darkness, and had her guarded while he went to the "world of light," *Mshuni Kushta*, to report this incident. Meanwhile, *Ruhayya* gave birth to her son *Ur*, a gigantic creature, who later became very curious about the whereabouts of his father.

So one day, his mother gave him a glass through which he could see his father's relatives living in the "world of light." This event made *Hibil Ziva* return to earth, while wearing his invisible talismanic face, so that *Ur* could not see him. He then captured *Ur* and made him a prisoner in the "Copper earth." His mother followed him and became his wife. Again, here one wonders who were those 360 regents that Mandaeans keep talking about, were they some visiting aliens?

In another Mandaean folk tale (as mentioned before on p. 220), once more appears *Ruhayya* with her son, trying to make Adam. But their Adam looked and walked like an ape, so they went to the "house of life" and told *melkis* about their failure. To fix the problem, *Hibil Ziva* was sent to earth bringing the soul, *Neshimta*, with him. However, the soul, horrified of Adam's appearance, declined to dwell in him. Later, after she was assured that while on earth she would be provided with all the delightful things that she had in the "world of light," she agreed to enter the body of Adam, which right away enabled Adam to stand erect and talk. Then *Hibil Ziva* taught him the secrets of life and also gave him the holy book of Mandaeans.

Children of Adam Son of Adam

In another Mandaean folk tales, Adam was made in the images of a *melki* called *Awature* and his son. They called him Adam *Boghra* or *Paghra*, which means the "physical body of Adam." Then, a rib was taken from Adam *Paghra*, and Hawwa (Eve) *Paghra* was made from it. The creation myth of Hawwa sounds particularly like a cloning procedure, and indeed a piece of bone would constitute very good tissue for this procedure. Ironically, the idea that Adam and Hawwa were created genetically identical would be in line with the genetic evidence that the humankind compared to other animal species is less polymorphic, that is, humans are genetically more similar to each other than most other animals are among their own kind. One way to achieve this level of genetic similarity would be starting a genetic line with a male and a female that are genetically very close to each other, as is suggested in this myth of Adam and Hawwa.

In any event, Adam *Paghra* and Hawwa *Paghra* had several children. One of them, called Adam son of Adam, took a wife among the *Shiviahi*, and that was supposedly the way our world got populated with humans. According to Mandaean folklore, the *Shiviahi* were people who had black skins and inhabited the earth before the creation of Adam and Hawwa. But according to Mandaean legends, they, that is, the Mandaeans, were not descended from the line of Adam son of Adam, but rather from the lines of other children of Adam and Hawwa who followed the teachings of their father.

So who were the *Shiviahi*? It is interesting that we usually entertain ourselves with the notion that when Adam and Eve were created they were the only human inhabitants on earth. Whereas according to many folk tales or other ancient traditions there were other human-like beings on earth, or at least those who had enough genetic compatibility with humans to interbreed with them. Were the *Shiviahi* modern-looking humans (*Homo sapiens*) with dark skin or were they Neanderthals (*divs*?). According to Mandaean tales, *Ruhayya*, the mother of *Ur*, was the *Shiviahi* who seduced Adam son of Adam so she could have her own race. It should be mentioned that the concepts of *divs* and demons also existed in other ancient texts, such as Avesta and Rig Veda, the scriptures of Zoroastrians and Hindus, respectively. Thus, it seems there is more to these creatures than just being part of the Middle Eastern folklore.

The Hidden Adam

Going back to the creation myth of Mandaeans, there is yet another legend of Adam. In their cosmology, everything in the physical world has a counterpart, a twin, in *Mshuni Kushta*, the "world of light," the "heavenly earth," which communicates with its earthly counterpart. They believe that after death, when the body is abandoned, the soul takes over the subtle body of its heavenly twin who consequently rises to a higher plane to turn into the body of light. While there, after further purification and judgment, the soul merges with the realm of light. Thus, in the "heavenly earth," there were the twins of Adam *Paghra* and Hawwa *Paghra* called Adam *Kasia* and Hawwa *Kasia*. To Mandaeans, Adam *Kasia*, the celestial Adam, is the archetype of perfect humanity and the soul of every human. He is also called the hidden Adam.

The Mandaean notion of beings living in the "heavenly earth" that are the counterparts of everything existing on earth is reminiscent of the concept of *faravashi*, the soul of believers in Zoroastrian religion (as discussed on p. 245). Likewise, according to the followers of this religion, every being including Ahura Mazda, the Supreme God, has its own *faravashi*. The concept of *faravashi* has a very deep root in Iranian culture and also boasts a continuous role in the development of Iranian philosophy. Despite the fact that the emergence of this concept dates back to a much earlier time than the era in which Mani lived, another aspect of *faravashi* called *farr* will be further discussed in the following section, in the light of Mani's concept of a divine spark in the bondage of the human body.

It is quite understandable that for many Western readers a detailed discussion of *farr*, a concept even enigmatic to most Iranians, will not sound that interesting. Nonetheless, if it is viewed not only as an Iranian concept, but also as a universal notion quite similar to "divine spark," then it is going to be worthwhile reading. Indeed, most people's understanding of "divine spark" is based mainly at a spiritual level; whereas *farr* goes one step further, bringing into spotlight a neglected portion of our mind which might be both divine and the source of human intellect and wisdom. Furthermore, without a prior deliberation about the enigmatic *farr*, the last chapter "Who Might We be?" would not make much sense.

The Enigmatic *Farr*

Pride of Ancient Kings

On the outskirts of the old city of Samarkand, presently part of Uzbekistan in Central Asia, and one of the oldest continuously inhabited cities in the world, going back 4,000 years, laid in ruins is the ancient city of Afrasyab, named after the legendary king of Turanian. All around this area is considered a sacred place of early humanity. Not far from here, in a cave called Teshik-Tash, located in the Baisun region, the complete skeleton of a Neanderthal child was discovered in the late 1930s. Then, more recently in 2003, in the Obi-Rakhmat rock shelter, which is located in Uzbekistan's old capital city of Tashkent, six teeth and over 150 cranial fragments were found. It seems that all these fragments belonged to some hominids (human-like beings) with features looking more like a mosaic of both archaic and modern humans, probably going back some 30,000 to 50,000 years.

Here indeed rests a land embedded with the forgotten stories of prehistoric ancestors of the ferocious Afrasyab and his wise minister Piran-e Viseh of nearby Khotan, better known today as Hotan in western China, an oasis town on the southern route of the ancient Silk Road. Afrasyab and the Iranian mythical kings, each accompanied by vast armies and heroic generals continuously fought each other during battles inundated with heartbreaking events. After all, they were all part of a big family. Iranians still name their daughters after Afrasyab's daughters, Manijeh and Farangiss; and after his land, Turan.

Shahnameh, the epic poetry book of the Iranian poet Ferdowsi, who wrote it some 1,000 years ago, is full of detailed stories of the lives of these long-gone warriors. Although in his book, he does not talk about Neanderthal bones, he informs us that he has gathered for us 30,000-year-old stories. How did Ferdowsi come up with such an ancient date? It is hard to say.

According to Ferdowsi, in ancient times (by some conservative estimates, about 5,000 years ago) there was an Iranian king named Fereydun who before dying divided his vast territory into three dominions and gave each piece to one of his three sons. The land he gave to his youngest son Iraj was called Iran, the land he gave to another son Turaj became Turan (the ancient Transoxina in Central Asia) and the land he gave to his eldest son Salm comprises the present-day Turkey.

Fereydun, by dividing up his land, unwittingly planted the seed of brotherly rivalry among his three sons and their young nations, in particular between Iranians and Turanians. But the real reason behind all that insane fighting, as Ferdowsi tells us, was not only because Fereydun gave Iran to his youngest son but something more important. He gave Iraj, his *farr-e izadi*, the *farr* that God had bestowed on him when he was a young man (see p. 240 for more about *izad*).

Apparently, many of the fights Afrasyab staged with Iranians were intended to regain the *farr* that Turanians shared with Iranians before Fereydun divided his land among his sons. So one wonders what was *farr-e izadi,* to be worthy of all that fighting? It should be reminded that, in ancient Iran, unlike some other ancient lands, there was never any indication that the king was a god in the form of a man. Gods were always gods, and humans were just mortal beings. But there was always a notion of kings being the representatives of gods on earth. In an earlier story in *Shahnameh*, Ferdowsi had spun another saga in the mystery of *farr-e izadi* and told us how Jamshid, a legendary Iranian king, lost his *farr*. Jamshid, a Solomonic figure in the possession of a magical golden ring and an ornate rod was a mythical Iranian king who appeared in the early stages of the formation of Iranian national identity. The ruins of royal Persepolis (a Greek name) in southern Iran, although much younger than his legend, are named after his throne, Takht-e Jamshid.

During his golden age, he reigned for 700 years, and there was no disease, no death among his people. It was a time when the weather was ideal, never either hot or cold, and every inhabitant looked young and joyful. To increase his territory to accommodate more people–since nobody was dying–in three stages, he added one-third to his land, each time by magically rubbing his golden ring in the ground. Then, to prepare for a long and harsh winter, he instructed *divs* to make him a wonderful garden divided into many sections with elaborate streets full of houses, and filled the garden with every kind of animal, including humans. Supposedly, in this garden the sun was always in the center of the sky and set only once a year.

He is attributed to be the first person who made wine and noticed its medicinal effects. Legend has it that through his *Jaam-e Jam* (wine cup), also called *Jaam-e Jahannama*, meaning the "cup that shows the universe," he could see the affairs of his entire dominion and also the whole universe. Supposedly, on the vertical side of this wine cup, there was a row of consecutive marks, each designating a different level of alcohol tolerance for different people. To this date still in Iran the word *paymaneh* refers to both the wine cup and measuring cup. However, the *payman* part of the word *paymaneh* means "covenant." The concept of gradual markings later became an allegory in Sufi mysticism for the level of spiritual purity each person would strive to attain. Not surprisingly, the notion of *Jaam-e Jam* as a metaphor for clairvoyance has always been a popular theme in Iranian Poetry. The Sufi master Hafez says it all when he serenades, "For years the heart has been desiring *Jaam-e Jam* from us, what holds within itself, it was begging from the stranger."

Some Western scholars have suggested that Jamshid's legendary cup might be the original Holy Grail, the renowned cup used by Jesus at the Last Supper. Also the designation of *Noruz*, the first day of spring and a day of festivities, as the first day of the New Year—a day still celebrated in Iran and some other countries in Central Asia—has been attributed to Jamshid. It is customarily called *Noruz-e Jamshidy*. Soon with all these things going for him, as is the nature of humankind, he let pride and arrogance prevail on him and called himself the creator of the world. Then, one day *farr-e izadi* left him and flew away like a bird called *Shahin* (falcon), which is a royal bird meaning "king-like." Subsequently, he became weak and was defeated and killed by his demonic enemy.

The Divine Right to Rule

Farr or *farnah* is derived from the Avestan word *khvarnah* (*khwarnag*) meaning "having the radiance of." *Farr-e izadi*, and a similar phrase, *farr*-e Kayani (*farr* of the mythical Iranian Kayanian dynasty), is poetically described as the glory of God, as I learned in elementary school, and politically described as the divine right to rule, which I learned just a few years ago. It seems that the concept of *farr-e izadi* goes back to a very ancient time in Iranian history or prehistory (before the development of writing), when the function of a ruler was defined as king-priest, as was the case for Fereydun, who as a king was endowed with *farr-e izadi*. However, by the time of the Zoroastrian's authority, or most likely during an earlier time, a caste system was integrated in Iranian society in which the king's divine sanction was stripped from him and he was officially placed as the head of the warrior class, a class below the high ranking priesthood class.

That was probably the reason the role of *mubed-e mubedan*, the high priest of the Zoroastrian religion became very prominent in the royal court of the Sassanid dynasty. Indeed, this dynasty was named after Sassan who was not a king, rather a revolutionary religious authority of the time who inspired the first king of the Sassanid era to revolt. After this, when the opportunity arose, the Iranian kings tried very hard to reclaim for themselves the ancient *farr-e izadi*, the divine sanction to rule.

The rivalry between Iranian kings and high priests to represent the divine glory resurfaced again after Shia Islam officially became the state religion of Iran during the Safavid Empire of Iran about five hundred years ago. That was the time when Iranian kings become the rulers of vast territories again and they started to realize the problems associated with being controlled by religious ayatollahs. The word *ayatollah* literally means the "one endowed with a sign of God."

Another episode of this rivalry showed up more recently, at the onset of the Iranian Islamic Revolution, when the concept of *velayat-e faqhiyeh*, roughly translated as the authority/regency of theologians, was revived by Ayatollah Khomaini to question the legitimacy of Iran's ruling monarchy of the time.

I remember as a kid back in the school days, I always had a problem understanding the meaning of *farr* as the "glory of God." I thought, what *is* this glory that seems to be the main attribute of God? Even as a kid, I knew there must be more to the meaning of *farr* than glory because "glory of God" by itself, as an abstract expression, sounded very complicated. The first mention of *farr* in *Shahnameh* was when God chose Kayumarth (see p. 262), who himself supposedly had some cosmic origin, to be his deputy on earth, as if giving him a divine mandate to rule, which made me wonder: to rule over whom? In Iranian mythology, Kayumarth is always mentioned as the first man and the first king. Nevertheless, that was the period in which the first pair of humans who supposedly later became the parents of the humankind was not even created.

Those who have studied *Shahnameh* know, however, that there is an indirect indication that at that time the earth was populated with mythical *divs* who, in my view, show many similarities with Neanderthals, another human race which became supposedly completely extinct some 30,000 to 50,000 years ago, after coexisting with *Homo sapiens*, our current human race, over thousands of years.

Ferdowsi describes Kayumarth's *farr* very poetically "like the full moon shining behind a cypress tree." This description of *farr* is only one of many in which the enigmatic *farr* has managed to play continuously a big part in the development of Iranian culture, albeit in different symbolic forms. Whether carved on monumental stones in ancient times or depicted on Iranian national flag of more recent bygone times, it has always remained as a core belief in the Iranian ideology system.

The Emblem of an Empire

Ironically, the *farr*'s most recognizable symbolic representation is also its most enigmatic one. As left behind in the ruins of palaces or other monumental structures built during the Achaemenid Empire of Iran, there is, in several places, a depiction, carved on rock, of an enigmatic winged human figure that was the emblem of the Achaemenid kings some 2,500 years ago, who ruled a vast territory spanning from Egypt to India. The emblem is commonly known as *Farvahar* or, in modern Farsi, *Farohar* or *Frohar*. It has two long and beautifully detailed wings, spread out symmetrically around the waist, with a layered tail attached to the wing in the middle.

It is prominently shown hovering over the head of Darius the Great, as if protecting the king of kings, like an angelic figure, as Darius is shown seated on a simple throne at a palace gateway in Persepolis. The same symbol was also depicted on the outside walls of the royal tombs of Achaemenid kings, which were large rooms, carved out of solid mountain rocks, high above the ground in Naghsh-e Rostam, near Persepolis, south of Iran.

The serene looking bearded man at the center of the emblem, dressed in an official robe, and coming out of a ring in the middle of the wing, holds a smaller ring in his hand and was originally thought by scholars to be the depiction of Ahura Mazda, the Supreme God. But now it is widely agreed to be either the *faravashi* (see p. 245) of the king as Zoroastrians believe, or a symbolic presentation of *farr* itself as suggested by some other scholars of ancient Iran. The winged disk alone with no human figure in the middle part is also in full view in several places in Persepolis. The origin of the composite emblem is not Iranian. The bust of the human figure is similar to Ashur, the Assyrian god that was depicted during a historically earlier era. Also the winged sun disk was similar to the one representing the sky god in Egypt, also going back to an earlier period.

It has been suggested that Achaemenids cleverly chose the bust of the Assyrian god and the winged disk of the Egyptian sky god, both of them popular symbols throughout the Middle East at the time, so that the newly conquered nations could easily identify the mighty superpower with their own deities.

But as will be discussed later, the concept of *farr* itself, that is, the glory of god or divine glory, had its origin from a much earlier time in the Zoroastrian religion or even earlier times.

As mentioned before, the original word for *farr* was the Avestan word *khvarnah*, indirectly implying that the concept of *farr* had its origin in Zoroaster's teachings. According to some scholars, however, his teachings were indeed a reform of Mazdian and Mithraism traditions, which were the religions of Iranians before they converted to Zoroastrianism. Later, to retain the loyalty of the followers of *Mithra*, it is said that Zoroaster added this deity to the pantheon of Zoroastrian deities and placed *Mithra* under the authority of Ahura Mazda. It should be mentioned that it is believed that the 25th of December, Christmas day, was originally the birth date of *Mithra* or somehow was related to this deity. It is said that this date was introduced to the West through Mithraism, which was a prevalent religion among Roman soldiers in the first few centuries AD. Among Orthodox Christians, Jesus' birth date is commonly celebrated on the 8th of January.

It has been argued that the concept of *farr* most likely goes back further in time when Iranians were worshiping *Mithra* who represented and protected the sun, as their principal deity. It is said that there is a passage in the Farvardin Yasht, a Zoroastrian text, that states *khvarnah* was originally associated with the deities *Mithra* and *Apam-Napat*. The latter deity later was morphed into *Anahita* (*Anahid*), the protector of waters. The association of *Mithra* with *farr* especially makes sense when noticing that the word for sun in Farsi, *khorshid*, is made up of two words: *khor*, the new version of *khvar*, as in *khvarnah* (*farr*) and *shid*, meaning "ray of light."

As in ancient times, it was customary to assign a flower to each deity, the sunflower became associated with *Mithra*, and lotus, the blossom of the water lily (called nilofar in Farsi), became associated with *Anahita*. It should be mentioned that

Mithra and *Anahita*'s Farsi version, *Nahid*, as well as Nilofar, are still common names for Iranian girls. As described by Abolala Soudavar, an Iranian scholar and *farr* expert, the association of *Mithra* and *Nahid* with *farr* stems from the ancient belief that while during the day *Mithra* was the protector of the sun and its rays (its *farr*), after sunset when the sun sinks into the sea, *Anahita*, the deity in charge of water, became the source of the *farr* during the nighttime.

It is known that during the Median (Medes) reign, which was an Iranian dynasty before the era of Achaemenid some 2,700 years ago, the sunflower, a symbol of *Mithra*, was the official emblem of the Iranian court. It is depicted as a wheel-like simple motif with 19 identical petals surrounding a small circle in the middle. It was the sign of Medes authority and has been suggested to be their representation of Mithraic *farr*, a sign of continued power and victory over enemies. It was believed that while *Mithra* could bestow *farr* or take it away, *Verethraghna*, the deity for victory was the one with most *farr*. This Zoroastrian deity, who is called *Bahram* in modern Farsi, is also a common name for Iranian boys.

Interestingly, later in the Achaemenid palaces of Persepolis, the Median symbol of authority, the sunflower motif, and an elaborate depiction of lotus, the flower of *Anahita*, were everywhere in full view. These two motifs, alone or combined together, usually in small size and repeated endlessly in a long row are the main border decoration of walls in the palaces of Persepolis. These rows are also seen, in some places, symmetrically surrounding the *Farvahar* motif, with or without the bearded man.

In Iranian mythology, *Mithra* was also associated with the ram and other sheep-like animals. Those who have lived in a mountainous area among animals could probably understand very well the meaning of this association. They would know that at dawn it is the mountain goat that jumps to the highest rock and then stands still, mesmerized by the rising sun. With this in mind, it is easy to see why in some ancient paintings uncovered in Central Asia, golden lambs were the ones carrying the throne of the king. Ferdowsi, in *Shahnameh,* also describes a scene where by the order of Kaykavoos, a mythical Kayanian king of Iran, golden lambs bring a throne to Rostam, the hero of *Shahnameh.* This great honor was bestowed on Rostam because he courageously defeated the *divs* who had imprisoned Kaykavoos and his army. Interestingly, as Ferdowsi tells us, while Kaykavoos was imprisoned he lost his *farr*-e Kayani and at the same time lost his eyesight, which in Farsi translates into the "light of eyesight." Later, Rostam by dropping three drops of the blood from the liver of Div-e Sefid, the leader of the *divs* of Mazandaran, cures his blindness and thus Kaykavoos regains his *farr.*

The Auspicious Bird

Despite many associations of *farr* with *Mithra* as stated above, it is unjustified to say that *farr* simply represented solar radiance and thus, the everlasting power of the sun. As mentioned earlier, Ferdowsi likened Kayumarth's *farr* with a full moon shining behind a cypress tree, or something equivalent to eyesight in the Kaykavoos story. It is more likely that *farr*, as its meaning "having the radiance of" implies, is associated first and foremost with light, whether from the sun or moon or even the human eye; and it was declared as such in a Zoroastrian prayer devoted to fire that identifies *farr* with the light of the sacred fire.

However, the enigma of *farr* is that in as much as it represents light, in Iranian mythology it is also strongly associated with birds. The word for "feather" in Farsi is *par*, which is also the root for the word "flying." *Par* sounds very similar to *farr*, considering that in many words "f" and "p" are used interchangeably in Farsi. Moreover, in most mythologies birds are the messengers between gods and humans.

As mentioned in the beginning of this section, Jamshid's *farr-e izadi* one day left him and flew away in the form of a bird similar to the falcon. The Avestan word for this bird is *varagna* (also called *varghan*). According to Avesta, the holy book of Zoroastrians, *varagna* was the carrier of *khvarnah* (*farr*). That is precisely why it is suggested that the winged disk depicted in the *Farvahar* emblem of the Achaemenid era represented the wings of this Avestan bird and thus, it is a symbol of *farr*.

In modern Farsi, this bird is called *shahin*, the same as "falcon" that literally means "king-like," since in ancient Iran the falcon and falconry were always associated with kings. There is a legend in Iranian folklore about a bird called *Homa-ye saadat*, meaning "auspicious bird," whose shadow on earth while flying would bring good luck and victory to those under it. The bird, *Homa*, a bird of prey, was known to be very wise and ate only bone. It would lift the bone off the ground and while sitting on a high cliff, would throw it off the cliff and then would eat its pieces, or maybe its marrow. Although these days in Iran nobody believes that a bird's shadow can bring happiness, the legend is still alive in the form of Homa and Homayoon, common names for Iranian girls and boys, respectively. Furthermore, *Homayoon*, also meaning "good omen," was frequently used as part of the royal titles of Iranian kings, even up to recent times.

The association of birds with the *farr* could be because they were always viewed as very wise creatures in Iranian folklore and mythology, probably reflecting the observation that some birds could live up to 100 years or even longer. There is a legendary bird in Iranian mythology called *Simorgh*, a very huge bird whose abode was in the middle of the sea. It had lived such a long life that it had seen all the events happening in the world and thus possessed the wisdom of all ages, the divine wisdom. In these stories, *Simorgh* typically gives three of her feathers to the protagonist of the story and tells him that whenever he is in need, to just burn one of them and she would show up right away to help him. Likewise, according to Zoroastrian texts, in Bahram Yasht, Ahura Mazda advises Zoroaster to look for a certain falcon with feathers having lots of *farr* that would bring strength and support to the person who holds the feather.

Intriguingly, the association of birds with *farr* in Iranian culture was not always due to their wings or feathers. In some monuments made after the Achaemenid era, *farr* seems to be depicted as a ring around the neck of certain birds. This ring, conceptually, could be similar to the ring held in the hand of the bearded man of the *Farvahar* symbol. His ring is described as a ring of investiture from God to a king.

Unending Resurrection

Although in these stories *farr* was mostly given to kings to legitimize and maintain their kingship over ordinary people, this most special gift from gods was meant to be neither everlasting nor hereditary. Victory over enemies would be a means to attain it or increase it, and defeat would decrease it, or it could be taken away altogether. According to Avesta, ordinary people are also eligible to receive this blessing from Ahura Mazda, and indeed they are encouraged to seek it, as the divine grace and guidance that it offers would bring success and advantage to regular people. In the current teachings of Zoroastrians, everybody is endowed with *farr*, but it is more radiant and powerful in those who are great in virtue.

After the end of the Achaemenid era, the *Farvahar* motif of a human figure in the midst of a winged sun disk disappeared completely from the royal palaces and other monuments that were later erected by other Iranian empires. It was only in the last century when Zoroastrians worldwide revived this symbol of light and wisdom and made it their official emblem. Nevertheless, the concept of divine glory or *farr* itself did not die out in Iran after the Achaemenid era. It just took other forms and appeared in new visual symbolisms. During the next two succeeding empires of Iran, Parthian and Sassanid, *farr* seems to appear in the form of various headgears worn by the kings of these two dynasties. Another new visual symbol during this time was a circular halo seen around the head of some of the kings of the Sassanid era about 15 centuries ago, as shown on several coins made during this era.

Some five centuries later when Islam had widely spread across Iran, around 1000 AD, Ferdowsi in his *Shahnameh* gave a new life to the old Zoroastrian concept of *farr-e izadi*. According to the Iranian scholar, Iraj Bashiri, Ferdowsi has mentioned 451 times the word *farr* in his 60,000-line epic poetry, which he wrote to revive Farsi and also the national identity of Iranians that were being compromised during the reign of Arab Muslim caliphs in Iran.

After the development of an Islamic esoteric school in Iran in the 12th century called Ishraqi, commonly translated as the School of Illumination, which was heavily influenced by Zoroastrian ideology, the light symbolism of *farr* reverted back to its original Mithraic sun motif. Several centuries later, in the 16th century, this version of *farr* was used for a new visual symbolism in the shape of a sunburst called *shamseh*, not in Iran but in a neighboring country. It was designed for the Muslim Mughal rulers of India, so it would appear behind their heads to represent their divine glory. Later, a similar sunburst motif shining behind a standing lion decorated the Iranian national flag for a long time before the advent of the Islamic Revolution in Iran.

Secluded in Common Names

Today *farr* lives on in the first names of about two million Iranians (out of a total population of about 70 million people). There are more than thirty common first names the first syllable (prefix) of which starts with *far*. Indeed there are as many names starting with *far* as the next three most common first syllables combined together.

The next three prefixes for common names are *par* meaning "feather," *shah* meaning "king," and *mah* meaning "moon." *Par,* as mentioned before, is very similar in pronunciation to *far* in Farsi. Although *par* by itself means "feather," as a prefix it is either related to flying, as in *parandeh* meaning "bird," and *parvaneh* meaning "butterfly"; or it is related to *pari*, meaning an "angelic-like being," as in *Parisa* meaning "like *pari*." Although *far* and *par* sound the same and have related meanings as discussed above, they are not the same. There is a glory associated with *farr*, which is lacking in *par*; *farr* is more like a shining feather or wing.

It should be mentioned that many Iranian names starting with the above-mentioned prefixes are part of a system of name configuration that utilizes preset modules. For example, *shid* meaning "ray of light," and *rokh* meaning "face" are two common name modules used only at the end of a name. Therefore, when *shid* comes after *far*, *mah*, and *beh* prefixes, the names become Farshid, Mahshid, and Behshid. Likewise with the *rokh* module, there are several common names like Farrokh, Mahrokh, and Shahrokh. Needless to say not all prefixes and end modules would traditionally end up together. Commonly only a set of preferred combinations of modules are used. Some specific modules together may sound better or make better sense.

Going back to *far*, by far the most common prefix for Iranian names of both genders, in my family, for example, my name is Farideh, my sister's name is Farnaz, my aunt's name is Farkhondeh, and there are also five second and third cousins, in my family, whose names start with *far*. Their names are: Faramarz, Fareshteh, Farnoosh, Farshid and Farrokh.

In modern Farsi, *farr* by itself means "glory," or "eminent" and "exalted." However, as a prefix, *far*, could mean its original meaning of glory and splendor, or it could mean something coming from above as in *fareshteh*, meaning "angel." Additionally, it could mean something happening in the future as in *farda*, meaning "tomorrow."

Interestingly, the association between wisdom and *farr* as was discussed above, when talking about the relationship between *farr* and birds, is also apparent in some words that start with *far*. For example, *Farzaneh*, which is a common name for girls, means "wise"; and one of the words used in Farsi for "insight" and "sagacity" is *farasat*. Common words for some mental activities also start with *far*, such as f*aragereftan* meaning "to learn," *faramooshy* meaning "forgetfulness," *farz* meaning "hypothesis," and *farhang* which means "knowledge" and "culture." So in general, it seems that *farr* implies a mental glory, as well as relating to ceremonial splendor.

Not every name starting with *far* is made up of two modules, however. For example, *fareshteh* meaning "angel," is derived from the root word *afrashtan* meaning "to raise"/ "to hoist." Another name, Farideh, is derived from *afaridan*, which means "to create." However, there is another word in Farsi, *afaridegar*, derived from the same root word *afaridan*, which is much more interesting to talk about because it means "God." Although in everyday conversation *Khoda* is mostly used to convey the word for god (any type of god, exactly as god is used in English), in our literal works in Farsi, there are three other words for God that are frequently used in its place. The two more common ones are *afaridegar* and *parvardegar*, and the third one is *dadar*, albeit less familiar to most people.

These three words are not the names of God but rather its different attributes. However, they are exclusively used to describe God. *Afaridegar* means the "Creator," *parvardegar* means the "one who fosters," which is a very unusual attribute for God. It is unusual because fostering and raising people require physical and personal attention, such as parents raising their kids. *Dadar* means the "one who rules with justice."

The word *afaridan* (to create) sounds similar to the English word *offer*. Considering that Farsi and English are both linguistically in the same family, that is, Indo-European languages, *afaridan* and offer might have similar meanings as well. If so, *afaridan* might imply a special kind of creation, that is, offering something special, such as *farr*, to an already existing entity. In other words, *afaridan* might simply mean the "process of offering *farr*."

Another interesting word that sounds very similar to *afaridan* (to create) is *afarin*. It literally means "bravo," "excellent." It is a very common word in Farsi, constantly used by Iranian parents and teachers alike to give encouragement to kids and grown-ups. Sometimes, I wonder whether *afarin* is related in its meaning to *afaridan*, the creation process.

Farr and an Imaginary Divine Experimentation

Kayumarth, as you may remember, was the first human being in the Mazdian cosmology who was entrusted with *Vohu Manah*'s divine wisdom. He was also according to *Shahnameh,* the first man on earth endowed with *farr-e izadi*. Imagine *Vohu Manah*, a cosmic being, transferring experimentally its divine wisdom, its *farr*, to Kayumarth, an earthly being. Then imagine again that it called its successful experiment *afarin*, "bravo," and made a covenant with Kayumarth to be his lord and protector. I wonder, was Kayumarth aware of the other terms of this covenant? Did he know anything about the actual experiment in which he received the *farr* talisman? Was he told for how long his descendents were going to be under the divine spell of *Vohu Manah*, the Wise Lord, the archangel with good essence? The word *parvardegar* (the one who fosters) as an attribute of God particularly fits this scenario very well. An artificial reproduction would require an agency to take care of every little detail of the experiment.

But, genetically speaking, before his transformation with the *farr*, who could Kayumarth be? His tentative identity will be discussed at the very end of this book in "Who Might We Be?" Although *Vohu Manah*'s experimentation, as mentioned above, is purely a speculative scenario, it is somehow reminiscent of the Mandaean story about how angels baptized Adam for the first time in the Jordan River and endowed him with a circlet of radiance, as described in the chapter "Who Were Adam and Eve?"

Now, let's revisit the creation of *Mashiya* and *Mashiyana*, the first human male and female pair, who according to the ancient Zoroastrian cosmology grew from a stalk of a rhubarb-like plant from the seed of Kayumarth. Could it be viewed as an artificial reproduction experiment, something like a sophisticated cloning process? Interestingly, according to some ancient Iranian folk tales, from that rhubarb-like plant, besides the first pair of humans, other human-like creatures grew as well. But they were all defective, such as having an ear or an eye in their chests.

Although these kinds of mutations have never been found in actuality in humans, some geneticists who work with fruit flies are quite familiar with these bizarre defects. They are known as homeotic mutations and are observed both naturally and through experimental procedures involving genetic manipulations. One typical example would be a leg emerging from a fly's head.

So why did *Vohu Manah*, the Wise Lord, want to offer his wisdom to Kayumarth? Hafez, a master Sufi poet, in one of his verses talks about God laying the treasure of his love in our heart. One may ask what does this verse have to do with *farr*? A treasure of love from God could symbolize many other objects in the allegorical language of Sufism. The second part of this verse makes it less ambiguous, when he refers to this treasure as the "shadow of fortune being spread on his ruined dwelling." For those familiar with Hafez vocabulary, the shadow of fortune is probably reminiscent of the shadow of Homa, the auspicious bird, which, as mentioned above, would bring fortune to the lucky ones laid beneath its shadow. And as discussed, Homa was one of the main visual symbols of *farr*.

In another poem, Hafez brings up the same theme again by saying, "The king of eternity (God) gave us the treasure of love-sickness since the time we laid our face upon this ruined abode." To lay face upon also means "to set out." Ruined abode, a common phrase in Hafez vocabulary, refers to earth as being the opposite of the perfect state in which heaven is always supposed to be. Thus, this verse can be translated as "since we started our lives on earth, God endowed us with the treasure of yearning for him."

Needless to say, *farr* and its many interpretations by various sources basically consist of some assumptions extrapolated from ancient Iranian folklore that, like all other myths, to many people may seem to be half imaginary, half metaphor. But at the same time, they could have been treasured memories transferred from one generation to the next. Through the ages, however, the original concepts of the old memories must have been so wrapped in successively changing languages that now many of them would sound unintelligible to us. It is indeed in the nature of change to leave behind few recognizable footprints.

Farr Symbolism in Sufi Poetry

School of Illumination

For those who have read the Quran, one of the most fascinating things about the Prophet Mohammad, as it is probably the case with all prophets, is how genuinely he admired and was in awe of God–something that is currently inconceivable for many people living in a world consumed with ideals of personal achievements. The esoteric movement of Sufism, a branch of Islam that originated in the 9th century AD by various sages living in the Middle East as part of a vast Islamic empire, in essence, dedicated itself to find a spiritual path for ordinary people to attain a spiritual state of pure love towards God similar to one that hallmarked the Prophet Mohammad's spirituality. Those who have read the poetry of great Sufi poets, such as Rumi and Hafez, have noticed the same profound love in their poetry for the God who has endowed the principles of nature with such astonishing colors and sounds, to say the least.

Early Sufis were not strict Muslims and relied on their own specific interpretations of the Quran, at the price of being severely punished by the religious authorities of the time. However, they all took their full understanding of the Quran as their starting point in reaching out for the universal love, which is sustaining the universe. Rather than believing in religious doctrines, they advocated a laborious Gnostic tradition, based on personal experience of the divine essence. Traditionally, a Sufi mystic is called *darvish*, who in Farsi has come to denote a person leading a very simple and content life almost to the point of looking poor to many materialistic outsiders.

There is a most beautiful and serene *sureh* (section) in the Quran called *Noor* (light). It likens God to the light and towards the end, it says, "…a lamp burning with the oil of an olive tree which is neither of the East nor the West, bursting into flame even though fire touch it not….And it is light upon light," (from *The Man of Light in Iranian Sufism* by Henry Corbin, 1971, as stated in its translated edition from French to English, 1994).

In the past, the beauty of this *sureh* inspired countless poets, and its enigmatic nature turned many thinkers into philosophers. The splendid phrase "light upon light" has invited many philosophical and theological speculations since its revelation some fourteen centuries ago. In Manichean terms it has been likened to the imprisoned divine spark after leaving the human corporal body and merging with its original light source, light upon light.

Yahya Sohrawardi, known widely as the Master of Illumination, was an Iranian philosopher of the 12th century. He essentially based his enormously influential esoteric School of Illumination (Ishraqi) on a very similar concept he called "light of lights." Ishraqi School, which shares some similarities with Sufism, nevertheless, was a distinctly unique order, almost like an elaborate mystical scientific discipline with its frequent references to the physical properties of light. At the center stage of Sohrawardi's philosophy was an inner rising sun, the sun of the heart. He described in much detail when and how it would rise from the spiritual Oriental direction of the soul. That is why his school of thought was called Ishraqi, meaning "coming from the Orient" and the "direction from which the physical sun rises."

Sohrawardi envisioned a universe of light, graded in endless declining degrees of intensity by the presence of numerous archangels and angels, occupying it from the zenith of totality of light to where it extinguishes in the abyss of total darkness of matter. This universe is accordingly maintained by the energy created from a dynamic interplay between the illuminated light and its reflected light in endless steps, analogous to the love given by a lover and the demur offered in return by her or his beloved. So much so that in effect it is the energy of love that enables the beings of this universe to move upward in this hierarchal realm of light.

Meanwhile, in constructing his universe, Sohrawardi borrowed heavily from the ancient Zoroastrian concepts. His universe of *noor al-anwar* (light of lights) resembles the seventh heaven of Zoroastrians, and he named his six archangels after the ones acknowledged by the Prophet Zoroaster. In addition, he brought to the center stage the old *farr*, emitting the light that radiates from the light of lights, brightening the light universe and penetrating deep down into the realm of matter, as a guiding source.

Unjustly, but as predicted, Sohrawardi became a martyr at the tender age of 38 for blasphemy against Islamic principles by the ruling of a Syrian royal court. Like Mani and numerous other martyrs before and after him, his crime was attempting an extensive philosophical elaboration on some existing religious dogmas.

A Revered Sufi Poet

I am one of those people who dislikes any translation of great Sufi poets like Hafez and Rumi. Recently, there have been several translations of these two poets with questionable outcomes. Rumi's poetry seems to have suffered much less during the conversion from one tongue to another, probably due to his unique style of extreme spiritual fluidity, sounding almost as if his poetry was written effortlessly.

On the other hand, Hafez' poetry is exceedingly technical, both in his choice of esoteric concepts and the way in which he meticulously sets each word next to other words with the utmost consideration, as if being a verse sculptor. There is such mastery in his choice of particular words that unavoidably it gets completely lost during any translation.

So with this in mind, there is a need for clarification with respect to why I committed myself to translate a few verses of his poetry in this section. While reading and studying his works, it is evident that Hafez was very preoccupied with every detail of human existence, whether physical or spiritual. In particular, he showed a keen interest in how the human became human, as a distinct animal.

He routinely describes some specific events or talks about unique characters involved in the creation or emergence of the human, as if he wanted to share with us some ancient secrets, all in allegorical terms, nevertheless. On the other hand, when one reads the interpretation of his poetry, either in Farsi or as translated in English, all his specific details regarding the human creation are categorically summed up as typical verses representing the human longing to return to her/his heavenly abode. So I thought maybe it is time to do a literal, word-by-word interpretation of some of his verses that pertain to the subject of the creation of humankind to better understand the thoughts of this revered mystic.

Hafez, the Iranian master poet of the 14th century, was born around 1325 and died at the age of 69. Hafez was not his real name, but rather a title officially given to those who could recite the Quran by heart. As he claims in one of his own verses, Hafez could recite the Quran in fourteen versions. Hafez actually means "guardian," referring to a very ancient and widespread tradition of guarding and preserving the religious teachings through memory, generation after generation.

In his poetry, Hafez sometimes draws on the triumphant light symbolism of *farr*, such as the sun, as expressed by Ishraqi School, or fire, the very symbol of Ahura Mazda, the Supreme God in Zoroastrianism. And some other times, in describing the *farr*, we find him reverting back to the nostalgic narrative of the Iranian Sufis; very much in line with the eternal ordeal of the Manichean's divine spark imprisoned in the corporal body of Adam, and longing to return to his heavenly abode.

But as every term in Sufism is expressed allegorically, for many—even some Iranians not familiar with the ancient Zoroastrian concepts—Hafez' poetry, with all its utterly beautiful verses, sounds mystifyingly ambiguous, meticulously veiling its secrets from the hearts and minds of his uninitiated readers.

All poetry works of Hafez are assembled in one book called divan, which is a general term for all poetry works. In his time poetry was not written down in books, rather meant to be read with accompanying music in royal courts or more often in taverns that were special houses frequented by various artists.

It is said that after his death, one of his close friends collected his poetry in one book, which later was acquired by an Indian king, and is still kept in India. Although Arabic is called the language of God because the Quran was written in Arabic, Farsi is called the language of angels. Indeed, it seems that Indians were quite aware of this allegory, and Farsi for several centuries was the official language of many Indian royal courts up to about two hundred years ago, before the British Empire established itself in India.

Compared to Rumi and Ferdowsi, Hafez was not a prolific poet, and there are only about 600 pieces of poetry left by him. Technically, each piece is called *ghazal*, which is a specific poetry format consisting of 7 to 10 lines written in verse; each line made of two parts. In *ghazal*, the two parts of the first line rhyme together and then the second part of each consequent line also rhymes with the first line. This particular configuration gives *ghazal* a unique poetic sound, and particularly in the heart and mind of Hafez, it reached its ultimate expression, whether sounding as calm as a sonata or as uplifting as a symphonic work. In Iran, the land of poetry, the so-called land of nightingale and flower, with some 700 national poets, he is called the Master of Word. Not surprisingly, he was quite aware of his poetic mastery and did not mind reminding us of this, once in a while, in his own verses.

In the realm of Hafez' poetry, there are many domains. In the spiritual domain dwell *Khoda* (God), archangel *Soroush* (the Holy Spirit of Zoroastrians), as well as *fareshteh* and *malak (malik)*, the general terms for angel and archangel, respectively; beautiful heavenly maidens called *hurian* and his dearest of all, the heavenly bird. Here, there are also various celestial gardens and heavens like Paradise.

Then, there is his real world, with the sky, usually referred to as "enameled dome," or "firmament"; the sun, moon (glorified in its various phases), and specific stars such as Saturn and Venus; and of course, there is his favorite sanctuary, the tavern and the alley of his beloved. In his real world there is light, there is fire, there is the mirror, and of course there is wine, his preferred real and spiritual drink; and then his most treasured belonging of all, his Quran, as told in many of his verses, his companion during the long hours of night and at dawn. Furthermore, besides the celestial and physical worlds, in his realm there is a hidden universe called *ghaib* (see p. 68 for a detailed description), not perceived by ordinary people, and the source of the knowledge and wisdom of mystics like him; and here, in this hidden universe, there are also luck and fate, things that even mystics don't know much about.

Then, there are his beloved mortal acquaintances: *pir-e moghan*, presumably the general name for the priests of a pre-Zoroastrian religious tradition in Iran, who taught him the secrets of the ages; *saghi*, the general name for beautiful bartenders residing in taverns who, in the words of Hafez, had long lustrous black hair, narcissus-like drunken eyes, and lips likened to the precious ruby. There are also his brethrens, the mystics, and of course his most beloved, *shakh-e nabat*, supposedly the nickname for a beautiful young woman he only met once, while delivering bread to a rich neighborhood in Shiraz, in southern Iran, his beautiful city of birth. Legend has it that in his youth, he was working as an apprentice to a bread maker when he met her.

As far as mythical and known prophets are concerned, his favorites seem to be Adam, Jamshid (see p. 277), Solomon, Khezr (by some accounts, the same as Hermes in Greek mythology, Elias of the Quran and Elijah of the Old Testament), Joseph of Canaan, and the one who is most highly regarded in all Gnostic traditions, Jesus.

His realm of poetry is also a place where there are beautiful flowers disregarding the pure love offered as songs by their adoring sparrows, moths annihilating themselves in the flames of candles burning till dawn, brooks to sit by and contemplate about our transient world, the morning breeze bringing smells from the alley of his beloved, and also another breeze, *baad-e Saba*, supposedly originating from the legendary Garden of Saba (Shiba), carrying news from far away places.

One of the most recurrent themes of Hafez' poetry is *farr-e izadi*, which he very often disguises in its many well-known symbolic forms as mentioned in the previous section. After all, using beautiful symbolism is the essence of poetry and a signature of great poets. His favorite *farr* symbolism, as presented in his many poems are: heavenly bird, sun, an auspicious bird called Homa (see p. 285), fire, Solomon ring, treasure trove, and of course *Jaam-e Jam* (a legendary cup of wine). He frequently uses this revered cup as a metaphor for divine wisdom either with its specific full name or simply as a cup of wine.

He also uses various metaphors for this cup, such as the mirror, since the clear and mesmerizing surface of the cup when filled with wine is itself perceived as reflective as a mirror. Nevertheless, it was wine, the legendary liquid content of *Jaam-e Jam* that was always used as the main allegory for divine wisdom in Iranian Sufi poetry, including the poetry of Hafez. He also quite often uses ruby-lips as a metaphor for the wine itself. So in his poetry, our imaginations are tested with metaphors within metaphors.

Archangelic Verses

In the following paragraphs, I will mention a few verses of Hafez' poetry pertaining to the concept of *farr*, just to show how this ancient Iranian notion has endured in Iranian culture as a living symbol of divine wisdom bestowed on the human soul both at its intellectual and spiritual levels. I understand that probably some scholars whose main jobs are interpretation of Hafez' poetry will not agree with my assessments. My main intention here, however, is to study his poetry from the vantage point of an evolutionary geneticist. That is, for a change, to take the events that he so adamantly is talking about literally and not merely spiritually. In this discussion, first the literal meaning of each poem will be given, and then as best as I can, I will attempt to describe his allegorical intention.

I will start with the verse that I talked about in a previous chapter about Hafez, the one that really mesmerized me as a kid, but understandably its meaning evaded me for a long time (see p. 69).

Last night I saw that archangels knocked at
the door of the tavern;
they molded the Adam's clay and touched it
with the wine cup.

Needless to say, as a child, I did not know that the wine cup was a reference to *farr*. So here it seems Hafez is suggesting to us that when Adam was being created, archangels bestowed on him the divine wisdom of *farr*. In the second line of this poem, he talks about the inhabitants of the heavenly realm drinking wine with him who was such a lowly and unworthy companion. Then in the third line, he utterly speaks his mind, albeit wrapped in a veil, and talks about something very enigmatic, a mysterious safekeeping load that was originally mentioned in the Quran.

The sky could not haul the safekeeping load,
to do the job, they drew lots on the name of crazy me.

There is another part in this poem that is equally enigmatic to me. Who were those archangels? Were they the same as Zoroastrian archangels, such as *Vohu Manah*, who was originally the Lord of Wisdom of Mazdians? Wasn't God supposed to be the creator of Adam according to the Abrahamic religious traditions of the Middle East? The word *malaik*, which Hafez used in this verse, is the plural of *malak (malik)*, which has multiple meanings. Its original meaning seems to be a "divine being" like an angel or archangel, but it has also been described as "god" or "king." In another of his *ghazals*, Hafez talks more directly about the role the archangels played in human destiny.

Come beautiful bartender and bring the wine
that the heavenly maiden mixed with
the perfumes of archangels.

So here apparently Hafez is suggesting that indeed the divine wisdom represented by wine is part of the essence of an archangel. Those familiar with his poetry know that he used wine allegory of divine wisdom quite often. The following verse is but one of them.

Learn the manners of mystics and be generous,
since it is not such a great deed,
to be like an animal who would not drink wine
and would not become a human (*ensan*, see p. 270).

Another great Iranian poet, the Master of Sufism, Rumi, who was of Tajik descent (see p. 114), was born in 1207 and died about half a century before Hafez was born. He lived during the Mongol massive invasion throughout Asia and beyond and was greatly influenced by its brutality as shown in his Sufism philosophy. I will mention only one of his poetry verses here just to show how he and Hafez used the same wine allegory to talk about the creation of Adam.

On the day of alast, your being
drank the wine from your bountiful food overspread,
you are the lord of non-dimensionality,
don't be a slave of positional effects.

Alast is used quite often in Sufi poetry and refers to the presumed first day of creation in which a covenant was made between the Creator and Adam, when God asked Adam, "Am I not your creator?" and Adam replied, "Yes, you are." So here Rumi is referring to the day of human creation as the day when

Adam drank the wine of divine wisdom, and thus in essencc became divine himself.

One of the most famous verses where Hafez used the *Jaam-e Jam* (a legendary cup of wine) allegory of *farr* and divine wisdom, as mentioned before, is the following:

For years the heart had been desiring Jaam-e Jam from us,
what held within itself, it was begging from the stranger.

In another verse, he tells us about the nature of this venerated cup by saying:

The gem of Jaam-e Jam is from the quarry of another
universe, why are you asking for the clay of potters?

In the following two verses from one of his most famous *ghazals*, he describes for us how a mystic, a *pir-e moghan*, by seeing through a cup filled with clear and mirror-like wine was able to solve his problems.

I saw him joyful and smiling, with the wine
goblet in his hand,
looking at that mirror a hundred different ways,
I said: When did the sage give you this
see-through-universe cup,
he said: When it was making this enamel dome
(referring to the sky).

Although wine and *Jaam-e Jam* are the central themes of Hafez' poetry, generally speaking, his most memorable verses are those in which he uses the heavenly bird as the carrier of *farr* and divine wisdom to humans, as exemplified by the following verses.

I am the bird of the heavenly flower garden,
how to describe my loneliness,
to tell how I fell into this time-trap of accident,
I was an archangel and the upper heaven was my place,
it was Adam who brought me to this abbey
that was founded on ruins.

These two verses are among my most favorite pieces of Hafez' poetry. The last line is particularly very interesting in which the archangel and heavenly bird of divine wisdom are represented as the same entity. Also it directly shows that the archangel he is talking about is a different entity from God because it came down to earth for the sake of Adam, apparently endowing him with the divine wisdom.

In the following piece, Hafez is describing for us why it is so difficult for us, as the vessel of divine spark, to break up from our corporal body:

How I circle around the sky of the holy realm,
since in the configuration blueprint,
I was nailed down to the body.

So far all the verses of Hafez' poetry described here represented the Sufi poetry of Hafez. But there is also a Hafez deeply influenced by Ishraqi School (see p.294). The next three verses, each from a different gha*z*al, show how he used the light symbolism of *farr*, both as fire and sun. In the first verse, he tells us how one night he attained the realization of the sun of the heart, the *farr* that Sohrawardi was referring to:

They enraptured me by the radiance emanating
from my essence; they gave me the wine from
the cup of the manifestation of my own merits.

In the following verse he talks about the fire symbolism of *farr*. The priests he is talking about here were *moghan* who were the priests of a pre-Zoroastrian religious tradition in Iran in ancient times. It should be mentioned here that *moghan* is the same as Magi, well known as the three Magi in the Jesus nativity story.

I am much loved in the abbey of priests because
the fire that never dies, it is always in our heart.

And finally, there is the following verse in which Hafez tells us what we are meant to do after realizing, within us, our divine spark, the *farr*:

From now on, we are going to emanate light
from our hearts to the horizons, since we
reached the sun, and dust was over.

Here "dust" relates to the magical visual effect we all experience when seeing the sunrays through a small opening, where a blue streak of floating dust always separates us from the sun. Thus, the disappearance of dust in this verse represents the end of our innate doubts, after reaching the sun of the heart.

There was yet another ghazal by Hafez that I originally intended to discuss here in its entirety because it seemed to me as if it was his proclamation of human evolutionary origin, albeit not in a Darwinian sense. In Darwinian evolution, human intellect is believed to have increased due to its adaptive value. Nevertheless, there is no survival value what-so-ever for the evolution of our soul, our innate desire to be connected to a supernatural realm. Therefore, the human soul has been regarded as a byproduct of the evolution of the human body and its vital spirit. But in this piece, Hafez, through mystical allegories, has presented the developments of the human body and soul at the same level, depending on each other without overshadowing one another.

However, eventually I decided not to include this very complicated piece of poetry in this section because after reading a poor translation of it by another person, I was so dismayed with the utter nonsense that was meant to be its translation that I could not bring myself to see another half-baked translation of such a profound piece of poetry.

Needless to say, there are many Hafezes in many cultures, speaking many different languages, telling in many versions the same stories about how the human soul was entrusted to us through a divine agency. Understanding whether this agency existed at all, whether it was solely spiritual or physical in nature, is a subject deserving the meticulous scientific scrutiny. Sooner or later science has to deal with the wisdom of our ancient times.

Understandably, it would not be easy, neither for scientists nor for religious people to deal with this issue. Ideologically, both groups abhor viewing spirit and matter under the same light. But understanding the source of humanity should not be a matter of consensus. Indeed, knowledge, if it is truly the sun of the heart, whether gained through reasoning or revelation, will always find a way to shine through.

How to Define *Farr* at the Biological Level

The Sufi masters of the Middle Ages, who had acquired visionary knowledge, were able to define the soul at three distinct stages. The lower stage, first perceived in vision as black tar and later as a black cloud turning into dark blue, is represented by the human psyche, the ego, the passionate and commanding soul. According to Sufis, through repetitious praying, it will attain a higher level, the so-called blaming soul, perceived in vision as a great red sun. It is the self-conscious soul, the intellect that criticizes and censures itself.

It should be mentioned here that repetitious praying refers to a process during which the mind is deemed to reach an altered stage—very much like the outcome reached through the continuous back and forth movement of a pendulum in front of one's eyes, that is, mental hypnosis.

Through the next spiritual metamorphosis, the soul will be ready to realize its highest capacity, visualized by a most beautiful emerald green color. It is the pacified soul, the one referred by Sufis as the heart, the subtle center of light, the Sohrawardi's *farr* (see p. 295). At the same time, the third stage is the most perilous and hardest to achieve. It is said that when entering this stage, if it appears like a distant circle, then there will be a thousand stages separating the seeker from the reach of the pacified soul. From this moment the passage becomes very terrifying and if the novice were to get close to one of these circles, she/he would be set on fire.

The last description in particular may sound more like pure imagination to some, a segment from a dream, or even hallucination. A psychoanalyst would probably see it as a sign of a tormented soul. But for those who have been there and had a chance to experience it first hand, this most terrifying vision is as real as an old memory; and as removed from inter-personal struggles as a fetus dwelling in its mother's womb.

So if the pacified soul or its other metaphors, discussed before as divine spark and *farr*, does exist as witnessed by many mystics, how can we find out whether it has a biological basis or not? Or rather we may ask: how can we characterize its physical manifestation at the biological level? Is it a small piece of brain tissue hidden away from all regular nervous circuitry, accessible rarely and only by a small fraction of people when favorable conditions are met? Or is it a matter of differential wiring of some specific nerve fibers?

Whatever it is, if it is a biological entity, then it has to be defined at the DNA level as well; and that would be undoubtedly the hardest part because, by definition, it has to be an incorruptible piece of DNA. A type of DNA not allowed to undergo any change in its base sequence. After all, it is representing the divine spark, not amenable to evolution and anything undermining its pure state.

So would it be possible to possess such DNA in our genomic arsenal? A genome that is prone to random or maybe even non-random mutations all the time causing it to undergo change, albeit incredibly slowly. Although in our genome there are regions denoted as hot spots, meaning they show a high degree of variation from one genomic segment to another and also among different individuals, nobody has yet noticed a segment of DNA in our genome that is completely invariable.

But it should be mentioned that presently it is still considered a big task to compare some 3 to 3.5 billion DNA bases between numerous people to find out such a mysterious piece of DNA. Some recent DNA sequencing projects have pointed out to several conserved regions between the human and some other organisms. These pieces of DNA usually relate to either cellular household enzymes, sharing the same essential functions in a range of organisms, or surprisingly are part of the non-coding DNA that originally was not expected to show any DNA sequence conservation.

Furthermore, DNA base conservation does not necessarily mean that the conserved pieces never mutated in the first place. There are cellular mechanisms, such as DNA proofreading and DNA repair systems the main functions of which are to keep DNA devoid of any mutations. However, the mutated bases are not removed with 100 percent efficiency; therefore, DNA does change through time, albeit very slowly.

If such a mysterious piece of human DNA were ever to be found in the human genome, of course by no means would it be any indication that it had come from a divine source. But if we assume, just for the sake of argument, that this piece of DNA was indeed related to the divine spark, there is yet another requirement to consider. It should be present in multiple copies or show signs that at some time it indeed had many copies. I should mention here that some genes in the human genome show this peculiar characteristic, and thus it would not be an unusual prerequisite.

The reason for this requirement is that in many myths and narratives about Adam, the alleged forefather of the humankind, he was endowed with much greater wisdom and divinity than ordinary people today are supposedly in possession of. One-way to explain this disparity would be to assume that

Adam and Eve originally had many more copies of this hypothetical "divinity gene." But later their offspring or future generations gradually lost many of the original copies either through well-known genetic processes, or mating with other types of human-like beings, as has been frequently mentioned in the creation myths of the Middle Eastern people (see p. 273).

Needless to say, the foregoing scenario about the possible origin of human intellect and/or soul is purely speculative, and was brought up here solely to have a chance to contemplate about our current genetic understanding of this hypothetical concept. Although using the term "divine spark" suspiciously sounds like trying to present a religious point of view, as will be discussed in the next section "Who Might We Be?" it does not have to be so.

As discussed extensively in the chapter "Who Are We?" currently, evolutionary speaking, either at the DNA level or other biological levels, there is no understanding on how the intellectual capacity diverged so differently between the human and his evolutionary cousins, the great apes. On the other hand, although the shape of the human skull has evolved during the last half a million years, when it diverged from Neanderthals, the brain size itself has not increased drastically.

Furthermore, neither the shape of the human skull nor its brain size has evolved considerably during the last 200,000 years. At the same time, archeological findings point out that about 30,000 to 50,000 years ago, a sudden change appeared in the level of sophistication of human stone tools (see p. 181). This particular evidence has been used for suggesting a possible sudden rewiring of the human brain during that time, by as yet an unknown mechanism, resulting in a new human intellectual capacity, not available to the humankind at any previous time. So with all these bits of information, it seems rather unjustifiable to confuse the evolution of the human body with the emergence of human intellect.

On the other hand, the typical explanation given by the proponents of Darwinian selection is that during human evolution, those individuals who were "smarter" were better adapted to their changing environments and therefore, left more children, who in turn propagated their "smart" genes in the subsequent generations. If the present is any indication of the past, there is no evidence anywhere to verify this simplistic notion. Indeed, generally speaking, "smarter" people tend to leave fewer children than other citizens. Arguably, it has nothing to do with their adaptive fitness, rather it stems from their personal choices.

Indeed, if one asks any evolutionary geneticist about how the human evolved to acquire higher levels of consciousness, they could not and would not offer any definite answer. Whereas if one asks the same question of some other scientists who really do not know much about the fundamentals of evolutionary genetics, they would most probably talk about the only evolutionary force that they have heard about or read about in the popular press, that is, Darwinian selection, better known by evolutionary geneticists as directional selection (see p. 135 for other types of natural selection).

But if pressed for the likely genetic mechanism(s) involved in this unique transformation that has apparently happened only in one species in the long course of organic evolution, there is nothing more that they can bring to this discussion.

The reason for the general fascination with Darwinian selection is also its main hindrance. That is, it is a very broad and simple biological notion. Any gene or a combination of genes, or any associated phenotypic characteristic that enables an organism to leave more offspring behind, compared to its competitors, is said to be under natural selection.

In the absence of a considerable number of well-defined actual biological examples, either at the level of organism or DNA, to insist that Darwinian selection is the main driving force of evolution sounds just like another belief system. Indeed, among many non-religious educated people, the contemporary mindset is embedded with a general belief that Darwinian selection has already provided us with a framework on how human intellect could have been evolved.

Ironically, this worldview is based more on a strong trust in scientific establishments than anything else. Trusting that no matter how directional selection has to fit with other biological processes, eventually it will work out as predicted by Darwin some 150 years ago, and pontificated by others since then—just like a religious belief. Although it is a credible theory and some people can relate to it, especially, in a social context, that is, the concept of "survival of the fittest," it does not necessarily mean that it is indeed the main evolutionary force.

As an evolutionary geneticist, my view is that currently, in the absence of any viable alternatives, the evolution of the human body, with its associated vital spirit, the astonishing inner functioning of the brain that we share with some other primates, seems to be a separate legacy from the emergence of our remarkable intellect and our higher level of consciousness, whatever the involved mechanism(s) might be.

Human intellect with its companion, its inspiration, that is, the love to know, is a unique entity that, time and again, has been universally defined as a kind of divine spark. Having the same basic characteristics, albeit represented by many different metaphors, it is still a basic constituent of many different human cultures, as it has always been during our long history and prehistory. If the young Western scientific traditions have naively decided to ignore all that, it's their loss. The rest of the world sooner or later will hopefully realize that it does not have to live by the Western scientific conventions.

At the end, I wonder if the Prophet Mani were present now, with his sharp mind and masterful brush, how would he depict humanity in these early years of the 21st century? Could he notice any improvement on the state of the divine spark dwelling in us? The 20th century was kind to our bodies and harsh on our souls. The holy trinity of capitalism, that is, power, greed and deceit, have spellbound people, selling them the notion that there is no scientific basis in believing in the uniqueness of the human soul. We have been led to conclude for ourselves that it is nothing but a superstitious belief, an outdated medieval way of perceiving the world around us.

Meanwhile, in the holy name of capitalism, we have turned into heartless robots with barren mechanical souls, serving the great god of capitalism and its many adoring archangels. Mercy has lost its grace in the capitalistic scriptures, and we have found ourselves yet another time in the abyss of selfishness and indifference to others, as we have always been.

In the past, we butchered each other with swords and daggers like a herd of wild animals. Now, thanks to advanced technologies, we kill each other with a barrage of bullets from machine guns like a bunch of cool robots. So seemingly, scientific enlightenment and modernity through progress have not succeeded in changing us a bit, as far as how we would rather deal with each other.

Sadly enough, it appears that after being serenaded for quite a while with the lullaby praising the virtues of capitalism, the human soul has just entered an era of deep sleep. As in the old fables we were told as children, when the whole town would fall sleep under the spell of a powerful witch.

Who Might We Be?

It was before six in the morning but still dark outside and the sun was not coming out anytime soon. It was December 21, the longest night of the year and I could not sleep any longer. All night laying half-sleep, half-awake, I was thinking about an article I had read the night before about the results of a study surveying the mitochondrial DNA of several ethnic groups. I was wondering why, as a geneticist, I could not accept the results presented in the article, which was about the spread of early modern humans all over the world, and the subsequent emergence of humankind's diverse ethnic groups. I was wondering, why was I so doubtful of the basic assumptions of the authors of this paper in analyzing their data? Why were they making such simplified assumptions (see p. 165)? After all, we are talking about various evolutionary forces interacting with each other during hundreds of thousands of years, since the time when our distant ancestors started to roam around the globe.

I was so tired of thinking about these issues since the previous night. All I wanted that morning, before the sun would come up again was to know what does *amorzesh* in *khoda bia-*

morzatesh mean? This is a Farsi phrase uttered when mentioning or remembering a dead person. It figuratively means "may God forgive her/him." But it must mean much more than the simple act of forgiveness because we use this word exclusively for God. People can forgive but they never can *amorzesh* each other. Apparently, only God is capable of doing so.

So what could it be? Is it a fixed operation or an after-life evolving process, like the process of reincarnation, the core belief of Hindus and Buddhists relating to the passage of soul through many lives to get purified? And what is it that is supposed to get purified? Is it the whole soul or maybe just a portion of it? What is purification anyway? I thought about the cursed man of the Manicheans, upon the realization of the imprisonment of a divine spark in his physical body. I thought about his tormented soul. I remembered Adam, from *the first book of Adam and Eve*, going to the gate of Eden over and over again every day, asking the guards to let him in; and the stern God always telling him, "Don't be so impatient, go live your life and after a certain time we will let you in again."

What is this certain time? Is it the same as the reincarnation period? The Quran talks about this "certain time" all the time as well. Could this transient period of time be the same as *barzakh*, supposedly a time or a place of "deep sleep," according to Islam, which spans from the grave to Judgment Day? Is God's *amorzesh* supposed to happen after this certain time? Our poor mind! What has it ever done to deserve such a harsh treatment? Or maybe we, that is, our minds, are being subjected to a very long-term scientific study by some outer space creatures. I was wondering, was the human mind or its mythical *farr* entity (see p. 278) wired by some sky masters? Was *farr* an artificial and foreign genetic entity planted in us by them to keep track of us, to make us connected to them, to make us wonder about them? So who are we and who are they? Are they the biological gods of our mind?

I wished the sun would be here sooner. How much I had missed it since the day before! I got out of bed, went to the kitchen and looked in the recycle bin for the weather section of the previous day's newspaper. It said the sun would rise at 7:21 am sharp. Being assured that the sun would rise soon, I sat at the kitchen table, looking at the dark sky through the window. It was sparkling with endless bright stars, reminding me that the real meaning of deity is "shining."

Then I wondered, what if some 30,000 years ago, in the land of *divs* (see p. 204) that we would one day dig out their bones and call them Neanderthals, a big shining vessel descended from the big blue sky and landed quietly at dusk. Maybe it came from one of the moons of Saturn or maybe from even more faraway places.

One by one, the *divs*, who were roaming around, surrounded the spaceship, angrily throwing stones at it. A majestic alien, as bright as the full moon, stepped out, looked around and courageously headed towards the bewildered *divs*. Several other aliens followed it. Maybe it was *Vohu Manah* (see p. 266), probably the same as *Nasa Qadmaya*, the First Man of the Manicheans (see p. 257).

Many years passed, many more shining vessels traveled back and forth between their planet and ours, bringing more luminescent aliens eager to contemplate further about the bewitched naïve-looking *divs*. And then one day, they all left and did not come back again, or maybe they came back every few thousand years, just to check around.

But by the time they left, one *div* was not a *div* anymore; it was subjected to many rounds of experiments, involving some genetic manipulations, something to do with its primitive brain. Maybe introducing some changes so it could be wired with its celestial masters. Maybe those celestial masters were some endangered aliens, running out of time, trying to embed their shining *farr* in the primitive brain of a Neanderthal.

So at the end, all those genetic experiments worked out and *Vohu Manah* or *Nasa Qadmaya* left part of its light in the body of a primitive *div* and called him Kayumarth, the mortal life (see p. 262). Then, in a beautiful garden, his cells were subjected to further experiments, for years to come, until eventually the first male and female modified *divs* were generated. And the two, who were named after *Vohu Manah*, human, were told to mate and propagate, to fill the earth with their own kind. They did and had many children and grandchildren who little by little mated with their cousin *divs*, the Neanderthals, and thus little by little lost their original *farr*, their divine investiture, until they ended up as we are, possessing only a tiny part of the light shield that the First Man of the Manicheans had left behind.

Then suddenly I become aware of the scream of the whistling kettle, bringing me back from the world of nonsense to our world of the senses, and I remembered that it was teatime, the best time of the morning.

But it is not easy to come back and live in our conventional world, especially for evolutionary geneticists, currently dwelling in the shadow of a powerful biotech establishment that has come to subtly dictate to us how to think when running experiments, and how to make conclusions when dealing with the results of those experiments.

By entertaining such a bizarre scenario for the sudden emergence of human intellect, I feel like a child sitting next to a big pond, dropping small pebbles, one after another, trying to agitate its still water, hoping to bring some circulation, some excitement to it again. Imagining the time when it was a rushing river bringing new life to the acres and acres of colorful wild flowers in the surrounding foothills.

I feel the same way about the field of evolutionary genetics now, in particular the field of human evolution. In the name of genomics and big time science, it has been in a stalemate for more than two decades. We hardly hear any new arguments about this subject these days. Where is everybody? Since when did genetics become a matter of consensus?

Now the efforts undertaken to understand the events that shaped our humanity in the past have diminished to the level of two fanatical competing camps, each claiming to know it all, religious fanatics vs. Darwinian fanatics. Rather than finding new approaches to re-evaluate our old assumptions, now it is all about who is right and who is wrong. Such an arrogant stance is not in the spirit of scientific discoveries. How disgusting it is when scientists turn into politicians. So, invoking an unusual and unexpected idea, such as the one I portrayed above, sounds better to me than no argument at all. Let's expand our thoughts and assumptions about the human origin. Rather than just accepting this or that assumption to get ourselves filled up momentarily, to be entertained intellectually, let's be patient.

From the rapid expansion and sophistication of human stone tools, starting some 30,000 to 50,000 years ago, it is clear that this period was crucial in the development of human mental capacity, and yet this is exactly the period that no change in the human brain size or shape could be detected in our fossil records. The one and a half century old theory of Darwinian selection (see p.135 for other types of natural selection) proposed to account for the evolution of human intellectual capacity did not take into account this new and crucial archeological finding.

Future DNA studies will undoubtedly be useful, but the kind of DNA understanding needed to address these kinds of issues will not be available at least for another 30 years or so. Although DNA technologies move very fast these days, figuring out the meanings of the results acquired through these advanced technologies is not as straightforward as scientists would like it to be. Understanding our DNA heritage as it relates to the recent increase in our mental capacity will be a major task of genetics for many years to come.

For me, there was yet another compelling reason to propose such a bizarre scenario. The history of scientific interest in human evolution goes back less than two centuries, but as far as our recorded history shows, for about five thousand years, different cultures have shown great interest and made speculations about this eternal quest. Since antiquity philosophers, poets, writers, sages, hermits and religious figures, have all contributed to our current knowledge of who we might be.

To the Western scientific establishment, these speculations are nothing but the products of human imagination and superstition. Much of it could be, but to brand all of it as nonsense, will only help to diminish the scientific standing among the common people. The majority of people living in this huge planet do not find any reason to see the world through the arrogant lenses of the Western scientific establishment. Even the current head of the Human Genome Project, Francis Collin, frequently mentions his belief in God.

The same way that many scientists, during the early stages of establishing physical sciences, dedicated their lives and endured many hardships to gain insight into the nature of things, the wisdom of old sages from different cultures did not come to them effortlessly, either. It was a long spiritual journey in itself. In most esoteric traditions, enlightenment is associated with the concept that the human soul, long separated from the divine, has the capacity to communicate with its divine origin, and thus receive revelations and attain enlightenment. In the Sufi Order of Islam, *dervishes* (sages) embark on a long path of self-discovery consisting of four journeys, laborious and often daunting stages in the way to divine reunion.

Very briefly stated, these passages are: first, leading a disciplined life; second, dedicating oneself to a path set by a *pir* (guru, an "old person" in its literal meaning); third, attaining enlightenment through experiencing annihilation in the divine; and fourth, experiencing persistence in the annihilation by becoming merged with the divine.

Despite our occasional, here and there, spiritual insights that we all experience throughout our lives, very few of us actually go through profound spiritual journeys manifesting themselves in revelations. The rest will experience some sort of mental journeys, which could be quite different for different people. And still, some will experience only a materialistic journey during their lives. Since at the young age of ten I decided to become a scientist, I did not give myself any chance to have any spiritual journey, but I have passed through four mental journeys so far: being raised in a traditional world, becoming attracted to the New World, living in the New World, and now being disillusioned by the New World, especially after my self-exile from the field of genetic research.

Now for me, the only legacy left of the New World is a sound, just a sound. The sound that I have heard so many times, in so many different places. But the one I remember the most is the one that I heard in Tehran's busy traffic one morning. It was some 33 years ago. At the time, I was an undergraduate student at the University of Tehran.

One morning as usual, I was driving to the university. My car radio was on playing some classical music. It was rush hour, and cars were everywhere. The traffic in the major street that I was driving through was moving slowly, and as soon as I approached a huge circular plaza, I heard Dvorak's ninth symphony, *From the New World*, from my radio reaching its climatic movement. I turned the sound volume of the radio to its highest and entered the plaza as if diving into a whirling ocean. A sea of moving cars and the blasting sound of Dvorak's symphonic music merged together, uplifting me to a euphoric state.

Many years later, when I went all the way from Stanford University to Iowa State University to do my PhD work in genetics, I had heard Dvorak's ninth symphony many more times on my car radio. But while in Iowa, not only could I hear it, I could see the music as well. In its slow movements, I could vividly see the vast Midwest plains, now all cornfields, under a huge blue sky; and in its fast movements, I could see the rushing, deep blue, Mississippi River passing by Davenport, Iowa, in its long journey down to the Gulf Coast.

I could see those images because I had read somewhere that Dvorak, who wrote the score for this piece while staying in New York, at the end of the 19th century, was actually inspired to write it during his visit to Iowa.

So now the once exuberant New World is just a sound in my head. And I wonder, did I change or was the New World transformed so fast into a purely materialistic world, becoming yet another hierarchal closed system, so relentlessly consuming our human experience?

I graduated, at the top of my class, from the University of Tehran and within a few months I came to the US to attend Stanford University and become a geneticist. My original goal was, after learning about the advanced topics in genetics, to work towards establishing a new genetic field reconciling the principles of genetics with the old Eastern view that the people born at about the same time would share certain characteristics with each other, the so-called astrology (see p. 91).

When thirty-one years ago, in 1975, I first arrived at the farm, as Stanford is known by the locals, approaching it from the serene Palm Drive, until reaching the vast green span of the Oval where, under a most beautiful blue sky, the sprawling monastery-like campus suddenly rose to full view, it was a place like no other place to me.

I left Stanford after receiving my master's degree in biology. And by the time I came back to San Francisco as a geneticist ten years later, in 1988, to work not on my own long-held original ideas, but rather on other people's projects, the flame of the New World was getting dimmer and dimmer in my heart and with it were gone all my feelings for a unique place called Stanford. I remember, a few years ago, when one day I was driving from San Francisco to Stanford to show the campus to my sister who was visiting me from Iran, I felt so distant from the Stanford of my youth that I told her, "I wish I could see this place through your eyes, as I saw it for the first time so many years ago."

I don't quite understand it myself, did the New World fail me or did I fail myself? But whatever happened, I have no regrets. From where I stand now, success or failure is just a matter of perspective, or as seen from an Iranian philosophical point of view, it is the realization of victory in defeat.

From Stanford to my garage, where I run a small business now, I have lost much by not reaching my scientific goals and potential. But at the same time, I have earned something that I never thought was important before, to have time all to myself, to think and rethink, from the Neanderthals to Adam, and from DNA to myth. Along the way, I also had time to ponder about all those who gave us these tales, many sages and many merchants, from the merchants of DNA to the merchants of soul.

Going back to the meaning of the little prayer that kept me awake before dawn, as told in the beginning of this chapter, the one that we Iranians offer for the soul of departed people, I think that now I know what it means. After linguistically analyzing it a little more, I do not think *amorzesh* means "may God forgive the soul." It is indeed referring to the literal meaning of forgiveness, the act of giving and offering, which in this case probably means "may God offer immortality to the departed soul." So it seems there is a long journey ahead of our little Manichean divine spark, our *farr-e izadi*, perhaps the same as the enigmatic safekeeping load mentioned by Sufi poet, Hafez, in one of his most celebrated poems, about a load that the sky could not haul, so it was given to the clueless human.

PART FOUR

WHAT NOW?

While driving home downhill on San Jose Avenue, it reminded me again how much I had disliked this area just a few years ago. But now it seemed okay to me. In front of me, I could see the panoramic view of several hills, dotted with box-like and brightly painted houses. From the right corner of my car windshield, I could see the BART train, high in the sky on its curved track, on a narrow bridge over 280 Free Way, pushing forward like a giant snake.

In no time, it was devoured by the rushing afternoon fog that had spread all over just in a few short minutes, floating in the air like thousands restless ghosts and melting together to obscure the horizon one more time. I felt the rushing fog even in my head, and that was okay, too. It was soothing in a way, like getting lost with no intention of being found. Out of the blue, I remembered one of Rumi's verses, "There is so much to know, but I do not know…" Then trying to forget my profound ignorance, I started to recite my favorite Rumi poem:

I am the servant of the moon, except moon,
not say anything;
with me except the talk of honey and sugar,
not say anything;
not say any talk of suffering, except the talk of treasure,
not say anything;
from this unawareness, not suffer, not say anything …

I wondered what made Rumi become Rumi.
Before long, a thick fog was everywhere.

Correction:

On page 235, paragraph 3, "17" should be "up to 10".